A Naturalist in Florida

A *N*aturalist in *F*lorida

A Celebration of Eden
Archie Carr

Edited by Marjorie Harris Carr ♦ Foreword by Edward O. Wilson

Yale University Press   New Haven and London

Archie Carr, 1909–1987

Published with assistance from the foundation
established in memory of Philip Hamilton McMillan of the
Class of 1894, Yale College.

Copyright © 1994 by Yale University.
All rights reserved.
This book may not be reproduced, in whole or in part, including illustrations, in any form (beyond that copying permitted by Sections 107 and 108 of the U.S. Copyright Law and except by reviewers for the public press), without written permission from the publishers.
Designed by Deborah Dutton.
Set in Sabon type by Tseng Information Systems, Durham, North Carolina.
Printed in the United States of America by Vail-Ballou Press, Binghamton, N.Y.

Library of Congress Cataloging-in-Publication Data
Carr, Archie Fairly, 1909—87
A naturalist in Florida : a celebration of Eden / Archie Carr ;
edited by Marjorie Harris Carr ; foreword by Edward O. Wilson.
p. cm.
Includes bibliographical references (p.   ) and index.
ISBN 0-300-05589-7 (cloth)
0-300-06854-9 (paper)
1. Natural history—Florida. I. Carr, Marjorie Harris. II. Title.
QH105.F6C35 1994
508.759—dc20   93-44919
CIP

A catalogue record for this book is available from the British Library.
The paper in this book meets the guidelines for permanence and durability of the Committee on Production Guidelines for Book Longevity of the Council on Library Resources.

10 9 8 7 6 5 4 3 2

To the next generation:
Jonathan, Jennifer, and Adam

Contents

Photographs from the Life and Times of Archie Carr follow page 90

Foreword by Edward O. Wilson ◆ ix

Preface ◆ xiii

Acknowledgments ◆ xvii

Maps ◆ xx

Wewa Pond ◆ 1

The Bird and the Behemoth ◆ 14

Jubilee ◆ 22

Sticky Heels ◆ 40

A Florida Picnic ◆ 47

All the Way Down upon the Suwannee River ◆ 51

Suwannee River Sturgeon ◆ 73

An Introduction to the Herpetology of Florida ◆ 84

Florida Vignettes ◆ 91

Tails of Lizards ◆ 100

Alligator Country ◆ 104

A Subjective Key to the Fishes of Alachua County, Florida ◆ 125

Hound Magic ◆ 139

Carnivorous Plants ◆ 148

The Cold-Blooded Fraternity ◆ 153

Living with an Alligator ◆ 160

The Moss Forest ◆ 165

In Praise of Snakes ◆ 187

The Landscapes of Florida ◆ 193

Armadillo Dilemma ◆ 204

Water Hyacinths ◆ 210

Triple-Clutchers ◆ 220

The Gulf-Island Cottonmouths ◆ 225

A Dubious Future ◆ 230

Eden Changes ◆ 236

Notes ◆ 245

Index ◆ 255

Foreword

In the Florida experienced by Archie Carr there were "yet remaining scenes inexpressibly admirable and pleasing," as his esteemed predecessor William Bartram put it two centuries earlier. From the Apalachicola bluffs to the Florida Keys, across a career spanning fifty years, Carr became a uniquely qualified guide through the dwindling wildlands of the state, the world authority on turtles, and the South's greatest regional naturalist of his generation. He also traveled far afield—to Central America, the West Indies, East Africa—and he graced literature with works about these regions: *High Jungles and Low, The Windward Road, Ulendo,* and *So Excellent a Fishe.* But the core of his experience was gained in Florida. As a young man he absorbed this native Eden into his bones and tried to transmit his feel of it to others.

I will flatter myself here by claiming a spiritual kinship with the master naturalist and turtler. He was born in Mobile, Alabama, a city that produces fewer scientists of his rank than Mongolia does water skiers. I was born in Birmingham but spent part of my childhood in Mobile, where relatives on my father's side had lived for generations. Carr was raised in Savannah, Georgia, where as a boy he kept a small zoo of pets, including lizards, snakes, and turtles. In close parallel I grew up in Mobile, Pensacola, and small towns in Alabama, with a brief sojourn in Washington, D.C.; a large focus of my life as a teenager was the collection and study of snakes, salamanders, and turtles. I believe that in my formative years I experienced the natural environment of the Deep South in much the way Archie did and that perhaps I can speak of his work with a certain emotional authority.

The wellsprings of his inspiration were the richness of the southeastern fauna and flora and the recognition that it was still poorly known. The exploration of that biota therefore held some of the same enchantment for the ambitious field biologist as did journeys to far-off tropical lands. There were (and are) some forty species of snakes, one of the largest local faunas of these reptiles in the world. When the southern Appalachians are included, the salamander fauna is the richest in the world. New species were still being discovered at a rapid rate when Archie and I were students, twenty years apart. On warm spring nights after heavy rains, you could (and can) hear choruses of frogs that rival those of the Amazon: the soft whistling of *Hyla avivoca,* the drumbeat of *Hyla gratiosa,* the trilling by *Pseudacris* cricket frogs (closely resembling the sound of a fingernail run along the teeth of a pocket comb), and the wailing of hundreds of spadefoot toads, the lament of hellbound souls.

And giants, real prodigies: bull sharks the size of canoes working their way up the Suwannee River and huge *Amphiuma* salamanders, called conger eels, two to three feet long, best hunted with nets in the shallow overflow of creeks after heavy rains. And Archie Carr's favorite, alligator snappers—North America's largest freshwater turtle, weighing up to a hundred pounds or more—together with real alligators and, south of Miami, saltwater crocodiles.

In this book Carr shows himself to be a Southern writer whose craft is disciplined by scientific knowledge. Most times he is a professor of biology, explaining such matters as the jubilees of Mobile Bay and the origin of the fishes of Lake Nicaragua. Then he adds a dash of the good old boy, a southern traditional, plain-spoken and wise. He also becomes a Hemingwayesque describer of people on the land, respectful of other cultures; and above all the naturalist so in love with the subject he must fight off the temptation of surplus detail. He is the enemy of the precious and artificial in language. He has no need of it; he has the real goods ready to deliver with scant embellishment. Yet the lyrical spirit is always there, and it breaks through pleasingly at intervals. Drawn by Florida's crystalline springs, he finds the best, Homosassa, "glowing in the dark hammock like a great, flat-cut jewel."

He fashions poetry out of sheer exuberance, and he animates natural history, like no other writer before him, in the vernacular names of the plants and animals of his chosen domain. They roll off the tongue, jolting, attention-getting, half science and half local color: gaff-topsail catfish, golden topminnows, pusselguts, hogchokers, bluegill bream, longnose gars, stumpknockers, spreading adders, gopher tortoises, chicken cooters, pignuts, mockernut hickories, pickerelweed, maidencane, striking cedars, and so on leisurely through the byways of Anglo-Saxon venery and folk botany: Archie Carr stock-in-trade.

No writer exceeds Carr in his ambidextrous handling of human and natural history. He flexes this talent in the classic essay on the Suwannee River, in which he blends history, geology, and biology—switching back and forth between the 1774 narrative of William Bartram and his own observations, matching localities and plants and animals across time.

Like all major writers, Carr can be read at two levels. His stories are to be enjoyed straight and simple as adventure. But, more deeply, he speaks for the generation that saw its perception of nature profoundly change. When he was young, parts of Florida and much of the tropics seemed a secure wilderness. By his middle age the wildlands had been transformed into threatened nature reserves. Humanity has conquered the world, unfortunately, and saturated it. The primary forests have mostly fallen; tourists stand and play in crowds around his beloved springs. We cannot go back. Archie Carr understood that. I believe that part of the reason he wrote was to remember the way the world was before and never will be again, to help us carry some of the wonder over to the other side. It was indeed a fine sight, to use an Archie Carr expression, a fine sight. Let us turn to his essays to see it again.

Edward O. Wilson

*P*reface

This is a collection of the writings of Archie Carr completed over a span of fifty years. The writings reflect his ardent love for the landscapes of Florida, his growing knowledge of the intricate relationships of living things, and his entrancing ability to see humor in an amazing number of unlikely situations.

Archie spent his early boyhood in Mobile, Alabama, where his father, "Parson," was pastor of the Government Street Presbyterian Church. The family had a big yard and dogs, and Parson, a tall, thin, handsome man, often would ride his bicycle to the edge of town to hunt wild turkeys. He was a very successful hunter and usually came back with a turkey hanging from his handlebars. Some of Archie's earliest memories of the out-of-doors were of fishing trips, near Mobile, in a rowboat with his father.

The Carrs moved to Fort Worth, Texas, for several years, and hunting trips taken during that time made a deep impression on the young Archie. Often two or three families would go by wagon to camp beside a river. Archie's family loved the natural world and all it had to offer—including quail and dove. It was while the family was in Texas that Archie became enthralled by turtles. He never got over it.

The third major church served by Dr. Carr was in Savannah, Georgia. Hunting and fishing again were the main avenues into wilderness, though sailing with his high school friends in the coastal waters gave an additional dimension to Archie's knowledge of the natural world.

Archie came to central Florida with his family in the 1930s. He had the extraordinary good fortune of majoring in zoology at the University of Florida while pioneer ecologists Dr. J. Speed Rogers, Dr. Harley B. Sherman, and Dr. Theodore H. Hubbell were setting the guidelines and standards for their graduate students. Their charge to understand, in great detail, the natural surroundings of each animal resulted in a grand array of profiles of Florida animals. Through their students alone, Rogers, Sherman, and Hubbell made an enormous contribution to the rapidly developing field of ecology. The utter reasonableness of this ecological approach was enthusiastically adopted by Archie. It was the only way to look at an animal, particularly if you were curious and felt a warm and deep affection for the other forms of life on earth.

In those days, the state of Florida was, in large part, undescribed by biologists; though many scientists visited the state—like the enthusiastic Albert Hazen Wright and Anna Allen Wright of Cornell University, who came each year to collect and photograph the frogs and toads. Each new graduate student in the zoology department adopted a different form of wildlife for study, and the enthusiasm of the group was unbounded. The natural areas around Gainesville were visited almost daily, and trips were made nearly every weekend to other parts of the state. The interaction of four or five young biologists—each with a different specialty—as they explored, say, the Econfina River or Pat's Island in the Big Scrub, gave each of them an extraordinary breadth of understanding of Florida wildlife in its natural setting. Over and over again they ranged the state, from Pensacola to Key West. In the beginning Archie was interested in fish, frogs, salamanders, snakes, lizards, alligators, and turtles. Later he specialized in turtles—but through the interests of his colleagues he was also familiar with the ways of mayflies, crane flies, crickets, spiders, water beetles, crayfish, round-tail muskrats, bobcats, and assorted birds.

Four years in the highlands of Honduras in the 1940s sharpened his eye for the relationships of wildlife and wilderness. Here there were new landscapes and new associations of animals to understand. But pine-clad hills are typical of Honduras—their national anthem proclaims their beauty—and pine-clad flatwoods are typical of Florida. The similarities and the differences enhanced Archie's comprehension of ecology.

On his return to Gainesville and the University of Florida, Archie taught ecology, at that time a new and innovative discipline. The assignment suited Archie. Every week the class would be taken to a particular "landscape" in Florida after Archie had made an advance trip to see what was there. Visits into the wilderness were a continuing activity throughout Archie's lifetime. While he was working on *The Everglades* for Time-Life he made many trips into that mysterious and embattled area. I joined him on most of his trips around the state, and they were always exciting, interesting, and fun. "Adventure is a state of mind," he once said.

We lived on a two-hundred-acre farm near Gainesville. Well, twenty-five acres was "farm." The rest was woods of several kinds, sinkholes, lakes, swamps, and marshes. This was Archie's home base for more than forty years. Every day he spent some time roaming about the place, and with his uncanny eye—or luck—he saw many wonderful things. He named our pond Wewa, and his familiarity with and love for the place is reflected in his writings.

In the late 1960s Archie started to write a book about Florida. By then he had published books about Africa (*Ulendo: Travels of a Naturalist In and Out of Africa* and *The Land and Wildlife of Africa,*) Central America (*High Jungles and Low*), reptiles (*The Reptiles*), the Everglades (*The Everglades*), and sea turtles (*So Excellent a Fishe* and *The Windward Road*). It was high time he wrote about Florida. I do not think I am biased when I claim that Archie knew more about Florida wildlife and wilderness than any other person, today or in times gone by. And yet the prospect of writing about modern Florida gave him pause. At that time Archie wrote:

> When I set out to write this book I immediately sensed a danger looming. It was that I was almost bound to fall into the trap of nostalgia and indignation, of turning this book into a diatribe against the passing of original Florida. Because to anyone who has known Florida as long as I have, and whose main interest in the place has been its wild landscapes and wild creatures, the losses have been the most spectacular events of the past three decades. Fifty years ago John K. Small wrote a little book about this passing. He called it *Eden to Sahara*. To Small, even in those days, the changes were so dismal he simply had to cry out against them. Thirty years passed, the changes continued, and then, twenty years ago Thomas Barbour found he could no longer stifle his emotion over the losses, and he wrote *That Vanishing Eden*. T. B. was a very good friend of mine. I absorbed from him a lot of sadness over losses even older than I have personally known; and since T. B.'s time the waning has gone on apace without relenting anywhere.
>
> So being a naturalist, living in the woods, and having the peculiar background I have, I am especially susceptible to the disease of bitterness over the ruin of Florida—over the partly aimless, partly avaricious ruin of unequaled natural riches of the most nearly tropical state. But in my case I decided simply, "What the hell, you cry the blues and soon nobody listens." And that made me see there was really no sense writing another vanishing Eden book at all. A garment-rending sort of book would simply not be read, and so be garment-rending in the dark, and a waste of time. The way to get my point across would be to talk mostly about what joy still remains in the Florida landscape and then just sneak in some factual tooth-gnashing every now and then when the readers might really be reading.

And actually, as great as the destruction has been, there is a great lot left. The organization of the old landscapes has been disrupted, but scraps of it remain, and if your eye is tuned to the country, any venture beyond the subdivisions will turn up good things to see.

What is really worth writing, then, is probably not how Florida once was or how it ought still to be—that is, not a nostalgic lament of things lost, or a jeremiad on the evil now afoot—but rather a catalogue of what remains of natural Florida for the discerning eye to see. So I decided to write a book on just the things in Florida that I know most about or that most interest me.

So here is Archie's book about Florida wildlife and wilderness. Enjoy it.

Marjorie Harris Carr

Acknowledgments

For the past two years it has been my great pleasure to have Beth Ramey to assist me in assembling and editing this collection of Archie's writings. She knew Archie, and so her work has been, in a sense, a labor of love. Beth and I have worked together on conservation matters for more than ten years, and I prize her many talents. My heartfelt thanks to Beth.

For the past three years my assistant in a long effort to save a Florida river, the Ocklawaha (a project of the Florida Defenders of the Environment), has been David Godfrey, a young man of exceptional abilities in the field of communications. Until he redrew the map for the endpiece for the reprint of Archie's book *High Jungles and Low* (University Press of Florida, 1992), he didn't know of his talents as a mapmaker. I am greatly indebted to him for the superb maps he conceived for this book. Our old and dear friend Larry Ogren—a turtle man and a naturalist who is sharp with the cartoonist's pen—very kindly drew the little sketches of fishes to illustrate Archie's outrageous "Subjective Key to the Fishes of Alachua County."

Throughout this period of editing I have harassed several people for corroborative information. I am particularly grateful to several members of the faculty of the University of Florida: Dr. Frank Nordlie, chairman, department of zoology; Dr. James Perran Ross, assistant scientist, Florida State Museum; Bruce Chappell, archivist for the Florida history collections, University of Florida Library; Drs. Donald W. Hall and James E. Lloyd, department of entomology and nematology. Dr. William T. Haller and Victor Ramey, Center for Aquatic Plants, have answered my questions with patience and forbearance. Dennis N. David, wildlife biologist and Statewide Alligator Management Program coordinator, and Paul Moler, biological administrator at the Florida Game and Fresh Water Fish Commission, have brought me up to date on alligators, crocodiles, and hard-backed cooters. All the Florida Park Service people have been enormously helpful. I particularly want to thank Susan H. Dougherty of the Homosassa

Springs State Wildlife Park and Kathy Nagler of the Ichetucknee Springs State Park for their cheerful assistance.

Most of the photographs in this book came from an album Archie began in 1921 when his family moved from Texas to Savannah, Georgia. For special pictures I want to thank our old friend William M. Partington, who always takes grand pictures (he never cuts off heads); William Stitt, a boyhood friend of Archie's; Lewis Berner, a colleague at the University of Florida; Herb Press, audiovisual media director of the University of Florida Information Services; and Jo Conner, a talented photographer. Over the years Jo has chronicled many of Archie's activities. Lisa and Carl Wattenbarger, owners of the Light Work Lab in Gainesville, have performed miracles in reproducing prints from old, small, faded photographs.

I am most grateful to John Colson, publisher of *Wildlife Conservation* (formerly *Animal Kingdom*); John C. Weiser, director, photography and research, Time-Life, Inc.; Duncan Barnes, editor, *Field and Stream*; Dr. Walda Metcalf, senior editor, University Press of Florida; and Michael Robbins, editor, *Audubon* magazine, for permission to reprint selected portions of books and articles written by Archie.

I thank Susan P. Urstadt, my agent and an old friend, who guided the manuscript into the capable hands of Yale University Press and Jean E. Thomson Black, science editor. It has been a great pleasure to work with Jean, and I'd love to do it again.

As a matter of fact, throughout this endeavor everybody has been enormously encouraging, and I thank them. I particularly want to thank our children and their wives—Mimi, Chuck and Gail, Stephen, Tom, David and Peggy—for their love, support, and assistance. Mimi's help with proofreading and with the preparation of the index was invaluable.

M. H. C.

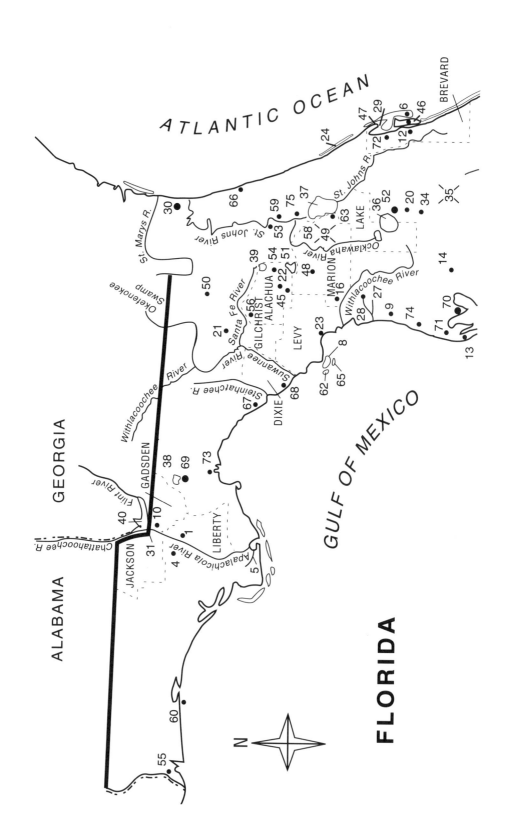

LEGEND

1 Apalachicola Ravines
2 Big Cypress Preserve
3 Big Pine Key
4 Blountstown
5 Brothers River
6 Cape Canaveral
7 Cape Sable
8 Cedar Key
9 Chassahowitzka Springs
10 Chattahoochee
11 Chokoloskee
12 Cocoa
13 Crystal Beach
14 Crystal Springs
15 Delray
16 Dunnellon
17 Everglades Park
18 Fakahatchee Strand
19 Fernandina Beach
20 Flamingo
21 Fort White
22 Gainesville
23 Gulf Hammock
24 Halifax River
25 Hialeah
26 Homestead
27 Homosassa River
28 Homosassa Springs
29 Indian River
30 Jacksonville
31 Jim Woodruff Dam
32 Jody's Spring
33 Key West
34 Kissimmee
35 Kissimmee Prairie
36 Lape Apopka
37 Lake George
38 Lake Jackson
39 Lake Santa Fe
40 Lake Seminole
41 Lake Worth
42 Lignum Vitae Key
43 Lower Matecumbe Key
44 Miami
45 Micanopy
46 Merritt Island
47 Mosquito Lagoon
48 Ocala
49 Ocala National Forest
50 Olustee
51 Orange Lake
52 Orlando
53 Palatka
54 Paynes Prairie
55 Pensacola
56 Poe Springs
57 Raccoon Point
58 Salt Springs
59 San Mateo
60 Santa Rosa Island
61 Sarasota
62 Sea Horse Key
63 Silver Glen Springs
64 Silver Springs
65 Snake Key
66 St. Augustine
67 Steinhatchee
68 Suwannee
69 Tallahassee
70 Tampa
71 Tarpon Springs
72 Titusville
73 Wakulla Springs
74 Weeki Wachee Springs
75 Welaka

SUWANNEE RIVER VALLEY

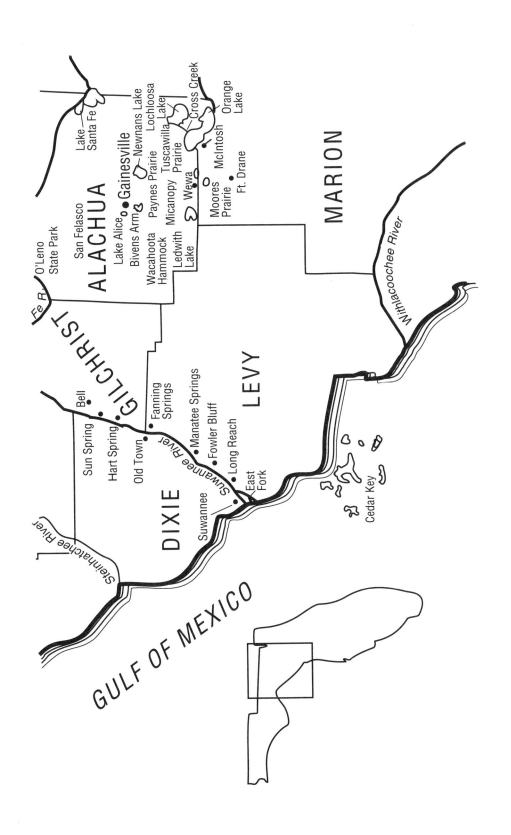

A Naturalist in Florida

*W*ewa *P*ond

There is a pond outside the southern window of our house, and watching its inhabitants has brightened our time beside it. At our breakfast table there are chairs on only three sides, so nobody has to turn his or her head more than ninety degrees to see what takes place on the pond.

The pond is not working alive with wildlife, but any time you look out across it you are likely to see some creature doing something worth attention.

One morning, for instance, I looked out over the rim of my coffee cup and saw deep waves running across a clear place in the bonnets. The waves were too big to be made by grebes cavorting or bass chasing shiners, and I watched expectantly to see which of the four most likely perpetrators of the disturbance—alligator, otter, anhinga, or cooting cooter—would appear at the surface. When the commotion subsided, no loglike gator showed itself, no gleaming otter body arched up seal-like above the surface. Instead, a bluegill bream a little bigger than my hand rose into view, poised a few inches above the water with no visible means of support, then continued to rise on the tip of a thin black stem that emerged beneath it. Then, ten inches above the water and flapping unhappily on its wispy pedestal, the thin-stemmed bluegill moved slowly away, rhythmically rocking back and forth on its stalk as it withdrew.

The unlikely apparition was a female anhinga, a snakebird, swimming as it does with its body under water and its prey held high above the surface—to cut down drag, I suppose, although my brother believes it is to weaken the fish by anoxia. In any case, when this snakebird reached a platform of floating bonnet roots she arched her neck over the edge of it

Adapted from *Animal Kingdom* (now *Wildlife Conservation,* a publication of the New York Zoological Society) 89, no. 3 (1986); *Animal Kingdom* 89, no. 5 (1986); *Animal Kingdom* 90, no. 1 (1987); and *Audubon* 85, no. 2 (1983) (copyright © 1983 by the National Audubon Society, 700 Broadway, New York, N.Y. 10003).

and, with much slithering, climbed onto the raft and began beating the fish against the roots.

This was a ritual I had witnessed often, and, as always, I found it hard to understand how a bird would know that the esophagus within its twig of a neck could possibly accommodate the spread of a mature copperbelly bream. I recalled a poor brown water snake with a body at least twice the diameter of the snakebird's neck that I found dead, the dorsal spines of a shellcracker bream projecting an inch through its body wall.

But snakebirds are evidently a bit ahead of brown water snakes in intellect, and it took this one only ten minutes or so to disarm its prey and begin the vast work of engulfing it. When the four-inch spread of the bluegill had finally slipped past the half-inch entrance to the bird's gullet and was halfway down her skinny neck—like a second body above the first—she slid into the water and swam across the pond. On reaching the far shore she climbed onto a snag and spread her wings akimbo to dry her feathers, which lack the oil that makes ducks waterproof, and to await the passing of the bream down her long esophagus.

Something is nearly always going on out on the pond. It has a serious defect, however. It lacks a name. When we bought the farm thirty-odd years ago, we couldn't find a name for the pond on any map of the region, old and new. Most of the old-time people said it just didn't have a real name, and nobody knew why. One man said he used to call it Curry Pond when he was a boy, but when we asked him who Mr. or Mrs. Curry was he said he didn't know and that maybe that wasn't the real name anyway.

So we naturally felt concerned and even a little insecure having a pond without a name in a region like northern Florida, where Indian, Spanish, and Cracker names for features of the land abound—where Wacahoota Hammock is just down the road and Tuscawilla is a short throw northeast of us, and where a little way to the southeast the Micanopy garrison fought the Seminoles in the Battle of Welika Pond. When we moved to our land there were only two indigenous names for anything on it. One was Smokey Hollow, at the eastern boundary, and the other was Stillhouse Pond, just across the pasture, where moonshine used to be made. Those are good names and we are grateful for them. But I'm growing uneasy over our pond's lack of a name, and I have decided to give it one.

So, I hereby name the body of water and the water meadows out in our yard Wewa Pond. I had to pore long over a Seminole dictionary to come up with that. The name recognizes no special arresting feature of the pond; Wewa just means water. But the word is Seminole, and it is short and easy to say—and the pond does have water in it. Or most of the time it does. When we first moved to its shore we were told that in the past one of the two sinkholes in the basin sometimes came unplugged, and the water would rush away down into the aquifer. On one occasion, old people said, this made a sucking sound audible a quarter of a mile away.

In spite of this unsettling tendency, Wewa is going to be the name of our pond, and if anybody objects, I suggest that he or she look up the Seminole names for more distinctive features of the pond than its water. For example, if you should set out to honor the most venerable spirit of the place—the alligator, which has been there longer than we have—you would saddle the pond with the unwieldy quadrisyllable Al-la-pat-taw. Few ponds have two residents as engaging as our great blue heron—Wak-ko-lot-ko, the Indians would call him—and the altogether fey young little blue heron, which Seminoles would speak of as Wak-ko-lot-ko-hi-lot-tee. The tree frogs that fill July nights with a din are Skin-cho-caw; and if you should decide to call the place after the marbled bullheads that abound in the pond, you would face Sar-sho-lo-ke-on-wah. It is that sort of difficulty you get into, looking for an appropriate name in Seminole. That is to say, few Seminole words come trippingly off the paleface tongue, and I am well satisfied with We-wa.

As in many other ponds and lakes in northern Florida, the usual slow pond-marsh-swamp-hammock succession is sometimes disrupted when the water suddenly drains away. There is little or no tectonic activity in our country—no earthquakes or volcanic eruptions ever in the memory of man. But the karst topography, as they call terrain such as ours, resting on soluble limestone, provides sporadic excitement when substantial bodies of water disappear or when the ground suddenly falls away as the roof of an underground cavern collapses. Both of these events occur from time to time in the limestone country of Florida, and both are remarkable occurrences.

The first I ever heard of such goings-on was during my first year at the University of Florida when a brand new 1931 Buick fell into a sinkhole. The owner telephoned Tarpon Springs for Greek sponge divers who could come up and fasten chains on it so that it could be hauled out. The divers went down but soon came back up, intimidated by the depth and darkness of the cluttered hole. So the Buick is down there yet, I judge—and only last year other cars were swallowed up when sinkholes suddenly formed. Within the city limits of Gainesville one of these dropped so far down that it was just left there and roofed over with concrete. A little later the premises of a car agency in Orlando collapsed, and two Porsches dropped out of sight. The agency got them out, however. They were very expensive.

Closely associated with such dramatic disruptions is the sudden draining away of bodies of water permanent enough to be on all existing maps of the area. In long periods of drought the level of the groundwater falls in the tubes and chimneys of the bedrock, and the solution ponds of the area—the ones connected with the underground system by vertical chimneys plugged with clay and debris—sometimes lose their plugs and go dry. This may happen only once in a generation or longer; when it does, it is a melancholy sight to see, with fish, snakes, and alligators crowded together in any deep place where water remains.

One of the most impressive of these ecologic cataclysms is the once-per-generation subsidence of the water of Alachua Lake, the vast solution plain known as Paynes Prairie, when it suddenly changes from an eight-thousand-acre area of navigable water to a few scattered pools and a ten-acre pond at the northern shore. When I was an undergraduate in the 1930s, Lake Jackson, near Tallahassee, sucked out for the first time in memory, and several of us drove over there to help rescue and resettle a couple of truckloads of turtles of half a dozen kinds.

It would probably be hard to overassess the effect that living in the soluble limestone country of Florida has had on the evolution and ecology of the animals and plants. The subsurface catacombs have greatly expanded the living space of the region. Some species have taken up their abode in the caves and water-filled conduits. Various kinds of crustaceans are wholly committed to life within the stone. All the crayfish are without eyes and pigment, and their legs are long and thin, like locomotor antennae.

From our well, which is two hundred feet deep, there sometimes come up little white crustaceans of the group the Crackers call gator fleas. One is *Asellus hobbsi;* the other is *Crangonyx hobbsi.* Both bear the name of my distinguished friend, Horton Hobbs, who used to spend a lot of time looking for new species of crayfishes in Florida caves. And finding them. One species, *Troglocambarus macleni,* is the most highly modified of all the cave crayfish, looking more like a ghostly relative of a scorpion than a proper crayfish. One of the most exciting moments of my life was when a friend handed me a jar containing a thitherto unknown genus of blind white salamander that came up out of a two-hundred-foot well. A moving experience for divers is entering a water-filled cave, playing a light about, and seeing the wraithlike crayfish teetering about the walls and ceilings in the stygian dark on their thin, sharp-footed legs.

To zoologists at the University of Florida it has always been an embarrassment—one that borders on insecurity—that there are no blind fish in the caves of the region, like those that grace subterranean waters of Cuba and Kentucky and many other places. Nobody knows the reason for this lack, but it must be associated with the interglacial rises in sea level that flooded Florida caves with seawater. Why the flooding failed to exclude the crustacean cavernicoles—or how they, and not fish, were able to reinvade the caves after the saltwater sank beneath the fresh—is not known, at least to me.

There are fish in some of the caves, but they are of kinds that occur in outside habitats as well. They are not blind or white or cave-adapted in any discernible way except by pre-adaptations generated by their ways of life outside. One of them is a catfish, the common yellow bullhead, a species that in Florida is often found in the depths of deep, dark sinkholes. Its habits must take it into the subterranean system, and once there the tactile sensitivity of its catfish barbels and its adjustments to the dim light of lime sinks allow it to survive in the dark. I don't know what adjustments to life in caves the chub minnow has made, beyond a tendency to feed on any edible particles that it encounters; but how it reaches the caves is easy to understand. It is strongly rheotactic, cruising about in little schools that usually orient upstream in the current. Moreover, it occurs most frequently in springs; and living in artesian springs and having a built-in tendency to swim upstream in little bands would, in Florida, be expected to take a fish into the honeycomb network of the aquifer. This is just what happens with the chub. I have seen chubs dug out of pipelike channels opened in solid limerock by a power shovel in a limestone pit five miles from any place where chubs lived at the surface. Throughout north-central Florida, yellow bullheads and red-eye chubs are, or used to be, found regularly in any cave leading down to groundwater. There can be no doubt whatever that they get there by underground routes.

Although the creatures that live in the caves and underground waters have conspicuous adaptations for subterranean life, the surface dwellers in limestone country—the inhabitants of the solution ponds, lakes, and prairies—are not so clearly marked by the karst environment. But to stand beside one of the last pools in the bed of a disappearing lake and see the predicament of its fauna is to feel sure that the stress of such a regimen must to some degree mold not just the ecological organization of the biota but its evolution as well.

When we first moved to the shores of Wewa Pond, the bottom had not fallen out of it for many decades, and it had a different look. There was little floating or emergent vegetation, and the whole eight-acre spread of water was filled with submerged plants—coontail moss, bladderwort, and cabomba. Our old alligator was present, even back then, and is probably the only inhabitant that has lived on from then till now. In the shallow water along the far shore a host of oversized soft-shelled turtles used to sun themselves, and there were more largemouth black bass than I have seen anywhere.

In those days, if I got home late and found I had forgotten the hamburger, say, I had only to grab a rod, jump in the canoe, and cast a weedless spoon into little openings in the coontail—and in fifteen minutes I could be back on shore scaling a couple of bass to feed to our ravenous offspring. Why bass were there in such numbers has remained a puzzle. I used to look for an answer in the stomachs of the ones we caught, but there was no explanation there; they were eating whatever small creature came along. Crayfish were what you mostly found.

Then came the cataclysm. The bottom fell out of the pond. In the big drought of 1956, when wells all up and down the Suwannee Valley went dry, the same thing happened to Wewa Pond. It took two days for the pond to shrink to a pool a hundred feet across. It was clear that the fish crowding into this dwindling refuge faced a dim future, so before they died we called the Florida Game and Fresh Water Fish Commission. When the commission workers arrived we dipped out fifteen hundred pounds of largemouth black bass, which the commission trucks carried down to a hospital in Lake County. I never learned who cleaned all of those fish. When the water returned the fauna came back, and it included all the species we had known before, but never again have the black bass abounded as before.

And there were other differences. For a pond to lose most of its water every once in a coon's age is obviously a drastic disruption of the ecologic regimen and very hard on the inhabitants of the place. In Wewa the loss of water is rarely complete, however, because the sinkhole that opens up is generally located higher in the basin than the deepest part of the pond's bottom, and a little water remains in a hole or two. Thus, when the sink plugs up again the animals that build back the fauna are usually the same species that were there before, though never in the same patterns of abundance and organization. Today, partly because of the sporadic loss of water and partly because the pond is really pretty small, no kind of big predator is abundant. A surprising number of creatures live solitary lives there or are represented, year after year, by a single pair. Wewa is by nature a one-alligator pond, for example; only one mature gator has ever lived there since we came. A big male comes in from somewhere to visit her every May, but when mating time is over he disappears.

Although this dearth of predators speaks badly for the productivity of the pond, it is in a way an advantage because it allows us to know some of the occupants as individuals.

One such individual that has drawn our attention for three years is a little blue heron. Little blue herons are apparently not widely known for striking eccentricities of behavior. One of their relatives, the reddish egret, is very clownish, instantly recognizable from a distance by its wild and seemingly uncalled for gawking and prancing while feeding. This ritualistic posturing is a racial trait, however, and its adaptive value is not obvious; it is probably useful in hidden ways. Anyway, all reddish egrets do it. The peculiarities of the little blue heron on Wewa, by contrast, are *individual* quirks, without evident survival value and, as far as my experience goes, restricted to this possibly retarded or mildly deranged young heron. It was this behavioral unorthodoxy that drew our attention to this particular heron and made it plain that we were watching the same one from day to day and month to month. With a less eccentric heron we might never have been sure of this.

L. B., as I will call him—giving both a name and an arbitrarily chosen sex for the sake of brevity of discussion—arrived on Wewa in April or May 1982. He was one of a group of twenty-odd young of the year, all in their juvenile white plumage. They knew how to fly but not to fish, and for a month or so they passed the days in some willow trees on a little island in the pond. They would leave the swamp just across the pasture where they were born and perch in the Wewa willows, stay there doing virtually nothing for several hours, then fly back to their natal swamp. We concluded that their parents must still be feeding them. I don't know that such protracted parental care is customary in the species, but otherwise it is hard to think why we never saw the young herons foraging—only sitting in a white cluster in the willow trees.

I'm pretty sure L. B. was one of that group, left behind when the others departed one day and never returned. Anyway, shortly after they went away we noticed a white little blue heron behaving strangely down at the pond. What it did was leap sporadically into the air and fly around a few feet above the water, repeatedly uttering the rasping croaks that herons make when startled or annoyed. It was never clear what evoked the complaints, which were usually cut off by L. B. diving headlong into the water, evidently after some real or imagined prey. That is to say, this prepubescent, newly flighted wading bird habitually dived for its food instead of taking it by stealth or ambush, and he usually made a lot of noise while searching for it.

This makes no sense, I realize. I tell of it only because L. B. has stayed on, turned blue, and stopped squalling and diving. He now passes his solitary days stalking about the pond and tweaking small creatures out of mats of water plants, like a proper heron. So even now his behavior is exceptional, because little blue herons are very gregarious, usually feeding or traveling in flocks, sometimes with white young and blue adults together.

I have no theory to explain the solitary life that L. B. leads. He adds character to the place, however, and we'll all watch closely to see what he does when the rookery in the swamp across the pasture is activated next month.

Another denizen of the pond that has drawn attention through the years is a great blue heron. He has a number of gripping traits. One of them is the basis for the name we gave him a few years ago: the Sport Fisherman. Most herons catch fish for nourishment. The great blue on Wewa does that too, but there was a time when he did it for sport as well. At least that is what he seemed to be doing, and for a couple of years we observed him closely. It may have been merely a sense of duty that made him catch fish he didn't want to eat, but I doubt it.

I realize that calling a bird a sport fisherman sounds anthropocentric. Actually, however, to believe that herons get fun out of fishing is only common sense. They are bound to enjoy it. In the present case, I think it was the heron who was anthropomorphic, not the people who watched him fish.

There was a bloom of catfish in Wewa at the time—big marbled bullheads nearly a foot long, with wide heads and menacing pectoral spines. They were not to be lightly preyed on by a predator prone to swallow its victims in one piece. In one of the wholly unexplained demographic changes that have occurred among the animals and plants in the pond, one age group of bullheads, at the time I am telling of, grew very abundant. They used to frequent the shallow water around the little island, and the lone great blue heron that lived on Wewa discovered this and for weeks did most of his foraging there. Over and over again we saw him walk quietly along the pond edge, stop and stand statuelike for a time, then suddenly strike out into the water before him and more often than not lift out a big black-and-silver catfish. He would then move ashore with his prey and start beating it against the ground. When the poor fish had quieted down, there was a spell of juggling to get it oriented head down, so the pectoral fins would fold harmlessly against its sides. Once this was accomplished the fish quickly disappeared.

Eventually these catfish must have grown beyond an ingestible threshold or else the heron just got glutted; because the time came when, more often than not, he would spear a fish, take it ashore, bash it against the ground a couple of times, regard it thoughtfully, then just walk away. Not away to rest or to brood but to catch another catfish. One day, after this had gone on for several weeks, I paddled over to the island and counted the remains of thirteen varyingly fresh corpses of marbled bullheads that the crows and coons and otters had failed to take away. It was then that we fell into the habit of calling our heron the Sport Fisherman.

After a while the baby catfish boom that had generated the bullhead baby bloom got even bigger in size or died off, and the heron stopped catching them. We stopped calling him the Sport Fisherman, and this pleased friends who always clean and eat what they catch in their sport fishing.

The great blue turned to other hunting and fishing, and some of his exploits have been dramatic. They sometimes get him into tight places and evoke fits of squawking that make you think that this time, for sure, the alligator has gotten hold of him. But then you see him wafting his thin form sedately away from the scene of whatever atrocity occurred, still squawking in outrage and offended dignity at the recollection.

The voice of a great blue heron is really one of the most outlandish sounds in American nature. A barred owl is even more impressive, but its night bellowings seem to be communication with other owls. The cries of the great blue appear to be motivated by gross overreactions to wrongs, and the wrongs usually seem quite minor. Even after years of hearing the Sport Fisherman when he goes into one of his raucous tirades you get the impression that things have gone totally to pieces out in the pond.

One repeated cause of these outbursts the heron brings on himself by trying to eat the alligator's offspring. Every year the alligator has young ones. For several months they stay together in a loosely aggregated pod, and the mother watches over them closely. The Sport Fisherman ought to be aware of this by now, but every hatching season, without fail, he stalks them, gets too close, and the old gator explodes out of her guard pool and almost catches him; and his grievous squawking can be heard all over our end of the county. On any summer day, if you hear violent thrashing and strangling croaks across the pond, you are sure to see white water over there and the heron flailing desperately away with the gator close behind him, half out of water, seemingly walking on her tail.

I should admit that there is no way of knowing whether the Wewa great blue has been the same individual all these years. Probably not. But the original Sport Fisherman of the big bullhead days was certainly a single bird. His behavior and techniques were too stereotyped to have been those of a series of solitary herons.

Recently there have been two great blue herons on the pond, and this is exceptional. This is the nesting season, and we are usually without any of their kind at this time. It is possible that the two may be a mated pair with a nest in nearby trees, but this would violate the usual colonial nesting habit of the species. In any case, one of these herons caught a bluegill even bigger than the one labored over by the snake bird I spoke of earlier. We watched to see whether he would lay it to rest on the island. But after a few minutes of stress he swallowed it, and he seems no worse for the experience.

This morning eight ducks—two shovelers and six blue-winged teal—

settled on Wewa Pond. They all arrived together and stayed only a little while, not feeding but resting quietly or paddling aimlessly among the bonnets. Once in a while they flew up and then quickly set down again. Their brief visit caused excitement in the house because for a decade Wewa has been almost devoid of migratory ducks. Ducks have suffered grievous depletion in Florida, and to watch this decline has been one of the penalties of living a long time in the same locality.

I have always liked ducks. I should say at the beginning, however, that my feeling for them has evolved as an incongruous blend of hunt-lust, gastronomy, and aesthetic appreciation. My earliest memories of live wild ducks is one of mallards sheering wisely away from our cunningly grouped patch of Montgomery Ward decoys, which were set out before dawn on a Texas lake by my father and me, then age seven, when a norther was on. In Texas style, the temperature had suddenly dropped to nineteen degrees Fahrenheit. We crouched by a little charcoal burner, and I remember the thrill and let-down as one flock after another veered into a climbing turn away from some flaw they saw in the sculpture or spread of our make-believe mallards, or from the shine of a high-brass shell overlooked on the floor of the boat. Other times my father would not come home empty-handed, and then it was that my mother took over and burdened me with lusts I would one day have to kick. She cooked ducks like an angel. So there was all that in my background; but then there came to be a love of wild ducks as biologic paragons and works of natural art, a feeling that has edged out the old hunting urge completely.

In thirty years no gun has been fired on Wewa Pond. The ducks that have come to it have been welcomed and regaled with corn. Besides being small and hidden in circling forest, Wewa is not, in most of its successional stages, prime habitat for either diving or puddling ducks. Nevertheless, every winter little flocks of ring-necks or blue-winged teal used to come in and stay for most of the season.

So the ducks of winter used to be one of the rewards of living on Wewa, and their passing has been one of the sad things there. Nobody who has moved to Florida during the past fifteen years can understand the grief of long-term residents over the loss. When I first came to the University of Florida as an undergraduate, U.S. Route 441 had just been built across the ten-thousand-acre marsh called Paynes Prairie. When you drove across it in January, there was never a time when ducks were not out ahead of your car, streaking east and west to scattered ponds and gator holes. This past winter I crossed the prairie every weekday and as always looked out for teal, but in a hundred round-trips I saw not a single duck of any kind.

This change has a strange, almost eerie feel to it. It does not merely reflect a decline of the prairie as habitat for ducks. It also means that for some reason our part of Florida has almost completely lost its share of the migratory ducks that used to come in from other parts of the continent. And for once a tragic thing happening to the wild world in Florida is not being brought about by Floridians. The state is losing wetlands—marshes, swamps, and solution prairies—to be sure, but there are still a lot left; and most of them are dismally empty of migratory ducks.

For a long time after the passing of my hunting years I clung to a shred of my addiction to ducks by watching and listening to their evening roosting flights over our pasture. There are wet prairies, lakes, and swamps all over the county, and ducks used to feed in some and sleep in others. Each species and little clique appeared to have its preferences. To watch them sort themselves out just before it got too dark to see their flight against the sky was a stirring way to pass a vesper half-hour. On clear, windless winter evenings, I would go out into the middle of the pasture, lie back-down on the Bahia grass, and look up into the deepening dark at an almost constant coming and going of ducks. They often crossed the field so low that on quiet nights you could hear the whisper, creak, or whistle of their wings. It was curious to look up and see a sky that all day long had been devoid of ducks, or marked only by high-flying V's, gradually come alive with the low-level commuting of teal, ring-necks, widgeons, wood ducks, and dusky ducks. The darker it got, the lower they all seemed to fly; and long after none could be seen at all, the sound of their wings would still come down, differing according to the species or the urgency of the travel: *cheek-cheek-cheek-cheek, shik-shik-shik-shik, queek-queek-queek.* . . . The evening duck flights used to be one of the reliable joys of those days, and today they are gone. The old pasture they crossed is out there still, just the same, and surrounded by the same array of ponds and lakes and prairies. But these winter evenings you can lie on your back till you get very cold, and no duck comes by in the violet sky above you.

If the same decline has spread widely, the migratory ducks of America are in a parlous state. I can find nobody who is able or willing to tell me whether the cause of their decline is lead-shot poisoning or loss of Canadian nesting grounds or short-stopping by northern hunt clubs or wildfowl management agencies.

There were times back in the 1960s when ducks grazed Wewa in fantastic abundance. In one of its ecologic moods the pond turned its resources to growing duckweed. There was a two-year period when no other floating vegetation was there, only tiny duckweeds of the four genera: *Lemna, Spirodela, Wolffia,* and *Wolffiella.* These covered the whole surface of the pond, and the ducks came after them in hundreds. Several times we estimated their numbers as at least a thousand, mostly widgeons and blue-winged teal; and one year the Christmas bird-count team recorded a thousand widgeons alone.

It was probably pure coincidence that during one of the biggest widgeon years hundreds of wood ducks also came. It wasn't duckweed that brought them. They were after the live oak acorns in a grove beside the pond. To the human palate the acorns of the live oak are far and away the most nearly edible in Florida, and in good acorn years—which come irregularly—the live oak hammocks are carpeted with them. At such times we have seen a couple hundred newly arrived wood ducks go ashore, crowd together in close ranks like dry fallen leaves, and move silently together across the yard and out into the live oak hammock beside the house. The wood ducks in our county are both resident and migratory. The migrants are fewer the past few years, but the residents—summer ducks we used to call them—are far more abundant than they were in my youth. The increase apparently reflects the decline of out-of-season hunting and of the numbers of irresponsible bearers of arms.

A memorable day back in the big duck years was the time the bobcat came to Wewa and ate one of our wood ducks. It was a golden day-after-Christmas. Up at the house the family had been lounging around in the overfed, more or less numb condition of the season, looking at ducks on the pond. The widgeons had come in several days before and had eaten through most of the six-acre patch of duckweed that had covered the water, and their numbers had dwindled to only a hundred or so. A lot of teal—mostly blue-winged but also a scattering of green-winged—were there; and a few ring-necks dived anxiously after what they could find on the bottom or sat still in patches of frogbit. Over in the knee-deep water of a buttonbush cove a hundred yards around the pond edge a swarm of wood ducks were whining, splashing, and making sporadic sallies out to harvest acorns under the long-limbed live oaks. After watching the ducks a while I left to go to the barn on an errand.

On the way over I saw fresh bobcat tracks in the sand road. It was a big cat, and the first hint I had of where he had gone and what he had in mind was the building-up of a curiously urgent chorus of whining wood ducks and squalling squirrels over around the pond.

When I got back to the house the people there were in a high state of excitement because a big bobcat had passed through the yard, with them watching from a window only a few yards away. He had walked past the window and moved down to the pond and then set out along the shore, alternately trotting and crouching or stopping to wigwag with his partial tail and peer out at the ducks on the water. As he passed along the shore, the wave of alarm that I had heard from over at the barn had gone along with him.

When they told me about the bobcat's visit, I went down to the pond and walked the way the bobcat had gone along the shore. The noise had abated by then, except for desultory squalling of squirrels, still stirred by their experience. The pond margin was marked with myriad splay-foot duck tracks on the mud, where the wood ducks had moved ashore to stuff themselves with acorns; and on this wide trail the bobcat had left his prints.

Under a big moss-draped live oak I saw how clearly the cat had known what he was about when he went stalking along the shore. On the floor of glossy little oak leaves there was a broad patch of feathers; they were feathers of a female wood duck. Fresh duck tracks led up to the site from the pond. Another trail, of wet cat-tracks, came up from the shore and entered a clump of dahoon holly that stood close by the carpet of feathers. So the story was clear. With his old cat stealth the bobcat had sneaked up into the holly patch, and from there one jump had closed the gap to where the duck was engrossed with the bonanza of acorns.

I felt a pang of regret that the duck had met her end. I was not, however, in a position to hold the bobcat's violence against him. As I said, I had eaten a great many wild ducks myself in my younger, undisciplined years.

The *B*ird and the *B*ehemoth

The herons were out among the cows when I got to the prairie. I saw them first, and then I saw the dredger working.

I slowed the car to creep along, irked by the sight of the dredger, bemused for the thousandth time by egrets and cows together. Then one of the herons got up and flew in under the swooping boom and lit on the pile of mud, all white lace in the slop and splatter. He was a snowy heron, and his coming in to stand there, though a small, ill-sorted thing to see, was for me a last chink stopped in a long daydream. It was a dream of birds and behemoths and of the smallness of the world, and its essence is this: the snowy heron remembers mastodons.

Not personally, of course. I don't know how long snowy herons live. Longer than a sparrow, no doubt—less long, perhaps, than a parrot. But racially the snowy is, I'll bet, a tie with times when, faunally speaking, Florida stood shoulder to shoulder with the Tanganyika of today. It is as plain as old bones or coprolites or rotten ivory in a Florida road-cut. Slim and impermanent as snowy herons seem, their race is old enough to recall a lot, and it does. The snowy remembers mastodons as clear as day.

Paynes Prairie is fifty square miles of level plain in north-central Florida let down in the hammock and pinelands south of Gainesville by collapse of the limestone bedrock. It drains partly into Orange Lake to the south and partly into a sinkhole at its northeast side. The sink used to clog up occasionally, and for years or decades the prairie would be under water. The people called it Alachua Lake in those times and ran steamboats on it.

Nowadays the prairie is mostly dry, with shallow ponds and patches of marsh where ancient gator holes have silted up but never disappeared, and with patches of Brahma cattle here and there out into the far spread of the plain, like antelope in Kenya. The prairie is about the best thing to

Adapted from Archie Carr, *Ulendo: Travels of a Naturalist In and Out of Africa* (New York: Knopf, 1964; Gainesville: University Press of Florida, 1993).

see on U.S. Route 441 from the Smoky Mountains to the Keys, though to tell why would be to digress badly. But everybody with any sense is crazy about the prairie. The cowboys who work there like it and tell with zest of unlikely creatures they see—a black panther was the last I heard of— and people fish for bowfins in the ditches. There used to be great vogue in snake catching on the prairie before the roadsides became a sanctuary. People from all around used to come and catch the snakes that sunned themselves along the road shoulders. When William Bartram was there the prairie wrought him up, and his prose about the place was borrowed by Coleridge for his poem "Kubla Khan." The prairie has changed since then, with all the wolves and the Indians gone. But still there are things to make a crossing worth your while, to make it, as I said, the best two miles in all the long road south from the mountains.

I live on one side of the prairie and work on the other side. I have crossed it a thousand times. Two thousand times. And always it is something more than getting to work or going home. I have seen the cranes dance there, and a swallow-tail kite, and, on the road during one crossing, 765 snakes. And there was an early morning in October that I remember. It was after a gossamer day, a day when the spiders go ballooning in the sky. Through all the afternoon before the spiders had been flying, young spiders and old of a number of kinds, ballooning to new places in the slow flood of a southeast wind. Some of them traveled no more than ten paces, riding the pull of their hair-thin threads for the space between two bushes. Some went by a thousand feet up, streaming off to Spain on jagged white ribbons like thirty feet of spun sugar against the sky. By nightfall the whole plain was covered with the silk of their landings. As far as you could see, the prairie was spread with a thin tissue of the dashed hopes and small triumphs of spiders, held up by the grass tips, draped over every buttonbush and willow.

We drove by in the early mist-hindered morning. The dew was down, and the drops formed strands of beads in acres of silken webbing. The fog had flaws in it here and there, and the sun coming through turned the plain all aglow, like a field of opals, and I slowed the car to look. Up toward the east from the road a Brahma bull stood in the edge of the sea of silk, and as I stopped on the shoulder across from where he was he raised his head to look our way. He was stern and high-horned and stood straight up from the forequarters, like an all-bull centaur. Suddenly the sunlight touched him, and my wife and I fell to beating at each other, each to be first to say: "Look at his horns." The old bull had gone grazing in the night, and now his horns were all cross-laced with silk picked up from the grass. He stood there with the sun rising behind him, and his horns were like a tall lyre strung with strings of seed pearls gathered in the mist and burning in the slanted light.

There is no telling the things you see on the prairie. To a taste not too dependent upon towns, there is always something, if only a new set of shades in the grass and sky or a round-tail muskrat bouncing across the blacktop or a string of teal running low with the clouds in the twilight in front of the winter wind. The prairie is a solid thing to hold to in a world all broken out with man. There is peace out there, and quiet to hear rails call and cranes bugling in the sky.

I slowed down, as I said at the start, just to watch the egrets with the cattle and to fret at the mess the dredge was making, sloshing about the old ditch, slinging muck about, scaring the cooters and congo eels. It was a big diesel dragline, a Lorraine 81. It was scooping fill for another pair of road lanes. It crouched in the mud on caterpillar tracks, and the steel boom that held the bucket up slanted away for sixty feet against the sky. In the shiny yellow cab a fat man snatched and shoved at a row of levers, barely able, it seemed to me, to keep up with the churning rhythm he was making. It was imposing, in the way of ponderous engines, the big, live-looking thing turning on the groaning bull gear, casting the brutal jaw, horsing up six-yard mouthfuls of spouting muck, and twisting to drop them on the growing fill. It was a gross, unlikely thing to see, a metal behemoth sloshed out to wreck the plain in vast quest of Mesozoic tubers.

It was that sort of fanciful thinking that slowed me down. But as I started to crawl on by, the egret left his cow and came flying in to the dragline. Straight in under the swinging tower it flew, and it lit on the piled new muck. I quickly looked up and down the road for a bump on a black and tan car that would mean a state trooper was coming. Seeing none, I pulled out onto the grass shoulder and shut the engine off.

I was only thirty feet from where the heron was, but it paid no heed to my being there. Its mind was all taken up with the fine things the dragline was spilling. Each time the bucket sucked out and rose, there seemed no way the heron could keep clear of the spilling mud. Each time the bucket dropped I looked for it to paste the bird flat with the splash. But always it jumped easily away and back again and fell to jabbing about and throwing its head up to juggle some squirming little animal and swallow it with hurried zest. Close in to the clamor and race of the engine, to the slap of cables and chains and chatter of drums and sheaves, it worked away, completely single-minded in its gleaning. Drag, hoist, twist, drop, twist, cast, drag; and the egret flapped in under the soaring bucket and took up the sad, succulent creatures from the mud, out of the midst of their disaster.

For a time I watched as if watching any unmeaning oddity. Then I caught a quick smell of half-burnt diesel fuel, and it took me back, the way odors sometimes do, to the deck of the *Piri Piri* on the Zambezi River and the smell in the air of a hot African afternoon, and to a flock of white herons standing on another plain with cows. Thinking about it in time and space like that, I saw all at once that a change had blurred the form of the dragline by the road. A sort of flesh seemed to be filling out the steel bones of the engine, and before my eyes it took a fleeting mammal form— not solid, skinbound shape, you know, but an eerie, momentary show of creature stuff partly condensed about the metal frame. You can't think how weird it was. It made me look hard; and after a bit I seemed to make out in the mist, still working away, still sloshing about in the ditch, the form of an old bull mastodon.

It was only for a moment. Then a car went by, headed south. The driver glanced the way I was looking but quickly turned back to visions of his own, to whatever draws the Yankees down to the end of Route 441. That made it plain that I was seeing untrustworthy things, and I looked back and, sure enough, the elephant had all ebbed away. The dredge was working for what it was, the motor straining at the drag chains and chattering through the turns, the steel mouth gnashing the muck to froth. But short as the stay of the elephant had been, it made sense of the heron's presence there.

When I moved to Florida the herons were wading birds. The cattle roamed the unfenced woods or sloshed about wet prairies—puh-raries, the Crackers called them—the marshes of maidencane, bonnets, and pickerelweed. The herons stayed in what you think of as heron habitat, in the shallow lakes and pond edges, and along the roadside ditches. They ate frogs and fishes there and little snakes and sirens. Even in those days there was a big cattle industry in Florida; but it was the hit-and-miss husbandry of the old Spaniards, profitable mainly because the stock was as Spartan as camels and land was cheap.

Then all at once the land began to change. A new sort of ranching grew up, with fences and purebred stock and planted pastures. New breeds were coddled in stumped-out, smoothed-over lands. The hammocks were cleared of brush and the palmettos were bulldozed away to make the flatwoods into parks. Patrician Angus, Hereford, and Indian bulls were sent out to serve the skittish Spanish she-stuff, and pretty soon, all over northern Florida, fat cows were being gentled on a new kind of tended lawn. By the thousands of acres the old rough land that no sane heron would be caught dead in was made into pangola parkland and clover savanna, into manmade pampa and veld of Bahia and napier grass. It was a landscape made over; and as strange as the change in the land itself was the change in the ways of the snowy heron.

To see what lured the herons ashore you have to understand that every Florida rancher, in spite of his dreams and all the courses he had in the College of Agriculture, is not likely ever to find himself raising cattle in pure culture. He will inevitably turn out to be a grasshopper husbandman as well. In fact, if he should stock his pastures with just the right number of cows, a number that eats grass exactly as fast as the grass replaces itself, a certain predictable yearly crop can be harvested. Of course, no rancher in his right mind works that way. He pieces out the winter diet of his cattle with protein and minerals and moves them to winter oats, and he does all he can to supplement the basic productivity of the pasture grass. But if only perennial grasses supported his cows, there would be in the pasture a certain fixed ratio of grass to meat.

Well, it's the same with his grasshopper culture. Let the insects move in and breed and live there with the cattle—and there is no way to keep them out—and the weight of the insect meat will be predictable too. The awful part is, it may not be a whole lot less than the weight of the beef. The herons of course know nothing of the rules of physiological ecology that make this so. But they know a good thing when they see it. And during the time between the two world wars, the snowy egret in northern Florida changed from merely a water bird to a seasonally insectivorous associate of cows.

When I first began to notice snowy egrets walking with cows the birds were with black cows mostly, and the two together were a fancy thing to see. I got into a great state of excitement; and knowing no ornithologist in those days, I canvassed the cattlemen of the county to see what they could tell me. Without exception they had noticed the hegira of the herons, and they dated it as "just lately"—lately being the early 1930s. When I went on to quiz them further, some said it was the coming of the Angus cattle that drew the birds ashore, some queer attraction of Angus black for egret white. But those with a less mystical cast of mind said it was not the blackness of the cows at all but the pastures smoothed out in the old rough hammock and palmetto land, the brand new bowling green laid out for a stick-legged wading bird to walk in. I looked through the bird books I could find and from them went to journals. Always the snowy was cited as a water bird—one who is more active in his hunting than the rest—but never as a cattle heron. Before the 1930s nobody spoke of snowy herons walking with cattle. By the end of the thirties they were observed doing it all over the place.

Looking back to those times you can see that several changes favored the hegira of the herons. These were the years when egrets were making their great comeback after plume hunting had reduced them almost to extinction at the turn of the century. And the new flocks were not surging back into the old Florida but into a land less fit for herons, with marshes and gladeland everywhere being drained and made into farms or real estate. Then there were the new crops of grazing grasshoppers on dry land, fed up to teeming tons on clear stands of planted grass, and there were placid cattle there to stir them out of hiding. And as important as any part of the new outlook was the change from the cluttered hammock and palmetto pinelands to lawns of short grass as wadeable to heron legs as water.

Traces of the sort of mind it takes to go ashore and consort with behemoths can be seen in the snowy heron's relatives. But in them the venture has the look of aberrant behavior, of a timid, unhappy straying from the comfort of the normal. The snowies, however, came out in confident flocks from the start. They emerged with unawed enthusiasm, as if loosed at last among joys once known and too long withheld from their bloodline. So nowadays the cattle quarter the mankept plains, the grasshoppers fly up, and the snowies snatch them out of the air. It was a rare thing they found, a feeding niche not occupied, a chance going begging. Any frog-spiking heron has the eye and the tools to tweak down a grasshopper out of the air; but only the snowy had the wit and the gall to go out and do it.

But *wit* and *gall*—what do they mean in a heron? What trait of mind was it, really, that singled out the snowy among his fellows and let him go out and use cows to harvest the new manna in the new landscape? Where did the flexibility come from? Why was the snowy so much the most ready when the new opportunity came along?

I think I know. I think they got inured to behemoths by walking with the fauna of the Pleistocene. Through millions of years Florida was spread with veld or tree savanna. Right there in the middle of Paynes Prairie itself there used to be creatures that would stand your hair on end. Pachyderms vaster than any now alive grazed the tall brakes or pruned the thin-spread trees. There were llamas and camels of half a dozen kinds; and bison and sloths and glyptodonts; bands of ancestral horses; and grazing tortoises as big as the bulls. And all these were scaring up grasshoppers in numbers bound to make a heron drool. Any heron going out among those big mammals—any small white bird able to make use of a glyptodont to flush his game—would have to have guts galore and a flexible outlook; but he would get victuals in volume.

Back among the ice ages and before, there must have been times, thousands of years at a stretch, when marshes and swamps went slowly dry, and frogs and puddle fishes grew scarce. At times like those the crotchbound kinds of herons could only mope and squabble about the dwindling water holes and starve there, or go away some place. But any heron strain with even a mite of extra flexibility would not need many generations to work out a way to live out in the grass.

Grasshoppers are hard to see, for a man at least, and I daresay for a heron too. They are colored all wrong to be seen, for one thing; and they have an unfair way of circling a stem of grass and sneering at you from the off side. But it is different when you walk down close to the nose of a cow. Out in front of a cow the grasshoppers are unable to use their cunning. They have to spew out into the clear like quail flushing ahead of a crazy setter. And for a fish-fast, frog-quick heron, picking them out of the air on the rise is no trick at all. The only hard part would be daring to move in close to the head of a creature a thousand times your size, the restringing of thin herons' nerves for consorting with behemoths—with cattle or mammoths or draglines.

There is no telling when the snowy's nerves were restrung—maybe as far back as the Pliocene, maybe further. In terms of geological time, climate has always been unsettled, and animals have changed with the climate or gone away or simply died. Again and again marsh has baked into adobe plain, tadpoles have withered, swamps have dried into forest and then into chaparral, and then through slow millennia have become swamps again. But even with all this going on the snowy would only have to shift his ways a little to survive—this way in the times when the fishes flourished, that way when the frogs became mummies in the cracking mud.

When the pterodactyls, the flying reptiles, mysteriously quit the world for good in late Cretaceous, there were aspirant bats to fit the living space they left. When dinosaurs dissolved away during the same calamitous times, mammals were on hand to take over their roles and skills and to think up many more besides. But in the more recent great extinction, that of the Ice Age grassland fauna, there has been only the most spurious replacement of what was lost. A whole life-form has dropped out of the old land-life structures. Throughout North America the whole grazing-browsing savanna community is gone or going. There is a rent-out space in the life-web where only a little while ago five kinds of elephants—and camels and horses, bison and shrub oxen, pronghorns and cervid deer—were making mammal landscapes that, you can see in even the dim evidence of bones, were the equal of any the world has known. It was in northern and central Florida that the great savanna fauna probably persisted the longest. Paleoecologists now say it might have held on down to no more than four thousand to eight thousand years ago. It has been no time at all since the animals were here when you think about how wholly they are gone, how empty of them the days are under the same sun and rain, how recently their horn flies dwindled, the condors mourned over the last cadavers, the dung beetles turned to quibbling over piles of rabbit pills.

And back at home you come upon a raging dragline with a wisp of a snowy heron there, dodging the cast and drop of the bucket as if only mammoth tusks were swinging—and what can it be but a sign of lost days and lost hosts that the genes of the bird remember?

*J*ubilee

The best cure for a complacent naturalist is to send him back into the field to look at animals. Being alive, wild creatures are unsettlingly prone to behave as they please. They upset one's preconceptions about them and impede the growth of natural history as an exact science. Take the freshwater jubilees, for instance. Nobody has the vaguest notion why they happen. The search for clues sends you figuring back and forth across the peninsula of Florida and down under the ground and even out into the sea. But before I move into the chain of byways that quest leads to, I should tell how jubilees got their name.

They say the word *jubilee* came into early English from the ancient Hebrew, where it meant a time of ceremonial trumpet-blowing to celebrate the freeing of slaves and the righting of other civil wrongs. I don't know whether the Israelis still use the word, but in the United States it is not nowadays often heard in serious speech except among southern black people or among white people speaking of blacks. Maybe it was the first part of the word sounding like *juba,* an old African name for a kind of dance, that caused black people to adopt it; or maybe it was the slave-freeing connotation of the term. Anyway, when I was young in Savannah, you might speak of a Cracker party as a blowout or a shindig, but one the blacks put on was more likely to be called a jubilee. That was in Georgia. When you got into the Gulf coast country between Mobile and Pensacola, jubilee turned out to mean something different.

Down in Daphne, Alabama, just a little way up the eastern shore of Mobile Bay from the Florida line, most days and nights go by in poetic quiet. But half a dozen times a summer, usually in the early morning hours, you will hear a sudden building up of excited talk outside, fast footsteps, and an unlikely racket of cars in the narrow streets that lead down to the Gulf. Above the growing din you can make out people yelling to one another, or to other people inside houses—or just yelling: "Jubilee! Jubilee!" They are not tolling in folk to any ordinary party. They are announcing a Daphne jubilee.

A jubilee in Daphne is like no other. It is a mass stranding of sea creatures of various kinds, which without evident cause have come teeming up out of the sea and into the shallows, as if moved by some craving to get out of the water and onto the land. The Daphne jubilees are stirring simply as natural phenomena, but to get their full feel you have to know the little town that for some reason is their epicenter. Summer is the season for jubilees. The earliest they ever come is April, and in April the old bayside section of town is a fit place for magic, though not for traffic and din. Daphne grew up in a hardwood hammock, and some of the old woods remain. Relict magnolias, hickories, and swoop-limbed live oaks hover over the little houses and let down a spattered mosaic of little patches of dark and light that dance or lie quiet according to the breeze. The yards and street sides are spread with a pastel spectrum of azalea bloom broken here and there by a snowy eruption of spirea. Some of the azalea plants are so ancient that they are little old gnarled trees; up higher the white dogwood flowers dance and blaze like four-rayed stars against the black-green magnolia leaves up higher still. In almost every yard wisteria falls in quiet lavender cascades, as if on the set of some wistful play about the antique South. On an April morning early, when there is no wind, or only a slow breeze, color and quiet coalesce so poignantly that the town seems bound to be only a dream.

Out on the shore beyond the bluff where the old streets end, the nostalgic feel is the same. The coast is eroding there, and cypress trees, knees, and ancient snags stand singly or in clumps along the sugar-white sand or out in the gentle lap of the sea itself. Up and down from the bluff-front, thin piers string seaward in teetering files—some still holding up plank walks—peopled at their distant ends by a dog barking into the mist or by a man and boy getting into a bateau or by a clump of women crabbing. Other piers, just the silver bones of piling set out by men long dead, now limp out through the mist toward where the oystershell sky and nickel of the bay merge indistinguishably. The dead piers are like some odd marine formation known only on this jubilee coast. The quiet feel of earlier times is strangely the same out on the shore and back along the seaside streets of the little town. Until the jubilee comes.

The coming of a jubilee is usually sensed by one or another of a few people who have the gift of seeing the signs in the sea and air. When the signs appear, they go down to the shore and wait. If the feeling has been authentic, flounders and crabs begin to gather in holes in ankle-deep bottom, and eels soon turn up where no eels were before. When the eels come, word quickly spreads inland, and people begin to move down to the beach with gigs, tubs, and flounder lights. Some of them build fires on the sand; some even set up camp on the shore. At the peak of a good jubilee there may be a thousand people on the Daphne beach happily harvesting an uncountable host of small edible animals of the bay.

A great many different kinds of bottom-dwelling fishes turn up at jubilees, but the main body of the gathering is composed of shoals of flounders; of crabs that congregate in clots or string out claw-in-claw in wild-looking chains along the shore; and of tiers of shrimp, which stay out a little farther than the crabs and fishes but still within easy reach of the minnow seines the people bring. At a good jubilee you can quickly fill a washtub with shrimp. You can gig a hundred flounders and fill the back of your pickup truck a foot deep in crabs. Or, if you prefer to concentrate for the sake of high gastronomy, it is easy to dip out a bushel of soft-shell crabs. In spite of their unnatural behavior, all these creatures have arrived, as far as anyone can see, in the best of health and spirits. The morale of the crabs is so high, in fact, that the males exploit the confusion and busily try to mount the softshell females.

All these diverse, unrelated creatures are simultaneously impelled by some mystic force to drop their affairs out in the bay and go to shore. Until 1960 nobody had any idea why they do it. People knew how to predict a jubilee with fair accuracy, but how the factors that foretold it were involved in producing the weird migration was a complete mystery. Then Harold Loesch, in a 1960 article in *Ecology,* proposed an explanation.[1] After looking over the jubilee shore carefully and talking with local people about the conditions of water and weather that seem to attend the gatherings, he suggested that a zone of water low in oxygen moves in from the depths of the bay and drives the animals before it. The oxygen is exhausted, he decided, by the decay of debris that washes out of the mouths of the rivers that empty nearby and accumulates in deep water just off shore. When protracted gentle easterly winds push the shore water seaward across the surface without building enough wave action to mix it with the deoxygenated water underneath, the zone of oxygen famine moves shoreward.

Just how each animal reacts when it encounters the zone of contact between the two kinds of water is not understood. But that is a problem of physiological ecology that never worries the jubilants or impairs their jubilation. People flock to Daphne from all up and down the coast, coming in hosts from Fairhope and from settlements far down the coast at Point Clear. People even telephone their relatives across the bay in Mobile, telling them to come over and reap the unsown harvest. From three to six days there is heavy traffic in Daphne. Then the creatures left uncaught move back out to sea where they came from, and the waterfront streets of the little town sink once again into their customary dreamlike quiet.

The highway signs at the city limits say "Daphne—The Jubilee Town." People there are proud of the curious distinction, and well may they be, because it is practically unique. Marine animals come ashore at other places too, but hardly ever do they come in shape for fish fries, crab boils, and gumbo parties. Down the coast at Fort Myers fishes sometimes strand by the thousands, but that is a very different kind of congregation. Those fishes are victims of the red tide plague. They arrive dead and inedible, and they soon send up a horrid stench that drives the tourists away. At Daphne, by contrast, the creatures come happily, firm-fleshed, and succulent, as if providentially sent to feed the people of the town.

So the term *jubilee,* as applied to unusual congregations of aquatic animals, originated on the eastern shore of Mobile Bay. But it seems appropriate for other congregations of aquatic animals that I have seen, and I aim to tell a bit about them.

One of these was the Wacahoota jubilee. I first saw this back in the time when the Kennedys were spreading about the land a new concern for physical fitness. Long walks had suddenly acquired a vogue, and one March day my son, David, then ten years old, set off on one of those man-building hikes with three companions. Their route took them down a dirt road through Wacahoota Hammock, a big expanse of hardwood forest that once was the scene of altercations between Seminole Indians and the soldiers of Fort Drane and Fort Micanopy. The forest is badly cut up nowadays, but some of it is still there, and the dirt road goes through a pleasant part of the remnant. The boys were aiming to take a twenty-mile hike. They came back after a little while, irked with one of their number who had grown tired six miles out and had hitchhiked home but excited over a vast congregation of little fishes in a roadside pool. David seemed uncommonly moved by what they had seen, and to quiet him we put some nets in the car and drove over to the place. Where a creek went under the road we stopped, got out of the car, and walked over to a pool on the downstream side of the road-fill. With David yelling at us to look, we looked, and we saw such a sight as I never saw before. There was a solid mass of thousands of little fishes, salamanders, and crayfish in the water. Five big culverts took the water of the creek under the road and dumped it into the pool. Peering into this, we not only could see no bottom, we could hardly see any water, the pool was so densely packed with animals.

There were tens of thousands of little animals there. They were of a number of different kinds, and they were to some extent segregated by species. One section of the mass was mainly killifish—golden topminnow and gambusias, mostly; another was layers of elassomas, or tiny perches; another was bushels of close-ranked crayfish. To cap the improbable fantasy of the scene, over at one edge of the pool scores of little sirens (two-legged aquatic salamanders, like black eels with feeble pairs of hands, external gills, and no legs at all) were undulating in an aimless ballet. For a zoologist who fancied himself steeped in the cold-blooded vertebrates of Florida, it was an incredible thing to see, a really unsettling thing; and I remember just standing there, ransacking the history of my close relationship with all those little creatures for any precedent or possible cause of their conduct. It was with something like resentment that I looked down at the immoderate gathering, as if creatures long known had proved suddenly not to have been known at all and so to have imposed upon me somehow.

That was my introduction to the Wacahoota jubilee. And as we stood there looking at it from the road-fill, I began simplemindedly mumbling, "A jubilee, a regular damned jubilee." And because there was no other word at hand for what was going on, the term stuck.

For several years we kept month-to-month check on the stream, and during that time four more jubilees took place. All occurred between February and April and, like the first, all lasted for several days. Because I was teaching a field course in ecology at the University of Florida at the time, the jubilee stream got an inordinate amount of attention from me and my classes. As time went by and no good explanation of the jubilees emerged, other professors from the zoology department set out separately to seek the cause. To this day no satisfactory answer has been proposed.

The jubilee creek flows north through the woods from a pond in a savanna known as Moore's Prairie, which now is a pasture. After passing under the road and continuing through the woods for a few hundred yards farther, the stream breaks up into distributaries, which enter Ledwith Lake, one of several big grassy prairies of the region. The distance between Moore's Prairie and Ledwith Lake is about a mile. Through much of its length the stream is fairly straight, reaches of it having been ditched out to drain Moore's Prairie and make a better pasture of it. The point where the creek goes under the road and makes the jubilee pool is about halfway along its length. The flow is conducted under the road through five two-foot culverts laid eighteen inches apart and set about a yard below the surface of the road. After every heavy rain the discharge is strong and turbulent, and the pool grows to roughly twenty feet in diameter, and much of its bottom of limestone cobbles is washed free of silt by the fast water.

Except at jubilee time there was nothing unusual about the stream. During times of normal water level it had the sparse vertebrate fauna characteristic of all brown-water creeks in North Florida hammocks. There were a few swamp darters and red-fin pickerel in it, gambusias in the quiet shallows, and maybe a crayfish every ten or twenty paces, if you looked hard for him. There were bronze bullfrogs along the banks. Otters, snapping turtles, and small alligators sporadically traveled the stream in their foraging between the two prairies. But the fauna is never abundant in the shaded, tea-colored hammock streams of Florida, and that of the jubilee creek was no exception—until some time in March or April when it rained for a day and a night and the water rose against the upstream bank of the fill, the culverts roared into the jubilee pool, another night went by, and suddenly there were animals in the pool by the thousands.

It was hard to think of a way to get good counts of the teeming jubilee animals. The water was too fast, the bottom too rocky, the different species too spottily and shiftingly spread about the basin, and the composition of the pool too changeable from day to day. But one day we made three short hauls with a little six-foot common-sense seine and laboriously counted out the catch. There were 7,505 animals in our bucket. Seventeen species of fishes were there, and in numbers far higher than normal for the stream. Of these, seven kinds made schools so dense that they hid the bottom or even boosted their own upper tiers above the surface of the water. These most abundant fishes were in their most usual order of abundance: mosquito fish, golden topminnow, pygmy sunfish, blue-spotted sunfish, flagfish, lined topminnow, and red-fin pickerel. The remaining ten species were less numerous, but there were still far too many to be in such habitat. The most abundant animal that night was the crayfish *Procambarus fallax*. We counted 2,084 of these. We took them home and ate them.

Besides the fishes and crayfish, on some jubilee days there were glass shrimp in translucent little bands, far more young sirens than one would dream of, and extraordinary sprinklings of the Louisiana newt and the striped mud eel (another two-legged amphibian of the genus *Pseudobranchus*). To add to the eerie confusion, some of the jubilees were attended by great gangs of big, restless red-bellied leeches. These behaved in a singular way. Instead of slinking leechlike along the bottom or among the rocks, they looped and undulated sportively about any free space left by the schools of backboned creatures, as if stirred to unnatural spirit and vigor by the plethora of potential hosts. But as far as I could see, none of them ever attached itself to any other animal.

To even the most detached or town-loving observer the Wacahoota jubilees were a wonder. To a zoologist who fancied himself familiar with the behavioral norms of the little animals involved, they were sensational. When word of a jubilee would spread to the University of Florida, small groups from the zoology department would instantly drop what they were doing, drive out to the hammock, and stand on the fill and discuss the phenomenon, trying earnestly to reason out a logical explanation for it. Through the years some very learned persons came, but little, as I said, ever came of their ponderings. Some of them would go away and get water-analysis apparatus and return to seek the cause of the jubilees in the chemistry or dissolved-gas concentration of the creek. All these efforts ever proved were that gradients of temperature, oxygen, or pollutants were not responsible. Some people used to try to work out the problem by pure deduction—some tried this without even leaving town. Hearing that the congregations occurred in a pool where culverts emptied, they suggested that the fast water issuing forth from the culverts simply blocked the upstream travel of the animals. When reminded that this left the motivation for the travel unexplained—that the jubilee fishes were mostly not kinds regularly found in the creek, much less in the numbers that turn up in the pool—these casual consultants might hopefully reply: "All right, they are in migration; maybe they just all take off on an upstream migration, and the culvert stops them." But merely labeling the travel "migration" doesn't explain it. And there still remains the problem of explaining what sets off a polyglot trek like that with salamanders, newts, leeches, glass shrimp, crayfish, and seventeen kinds of fishes all traveling in strange territory together. A prevalent reply was that the jubilee animals were all anadromous, or were of kinds that always migrate up streams to breed, and were going up the creek to lay their eggs. The idea, of course, quickly withers before the incongruity of all those creatures striking out on a breeding spree together, all getting struck at once by the same great idea: the crayfish, leeches, baby sirens, and half-grown pickerel; the live-bearing gambusias that produce a dozen young; and the sunfishes that build nests for twenty thousand eggs apiece—all of these going off in a happy band to breed. Still another suggestion I used to hear was that some unusual feeding opportunity might be involved. Maybe some of the smaller animals knew of abundant victuals up the creek somewhere and the bigger ones went along to prey on them. That thought broke down against the consistent emptiness of the stomachs of all the fishes, including even such strong predators as the pickerel and warmouth bass that were there.

And so the search went, and so it still goes; because no solution has been found. All that can be said with certainty is that the migrants come into the pool down toward Ledwith Lake and not from upstream toward Moore's Pond. This you can quickly prove by walking up and down from the road-fill for a way during the early stages of a jubilee. Upstream, the creek is almost lifeless, its natural sterility even greater than usual because of the rush of the flood water. Below the road, however, most of the jubilee species can be seen strung out in slowly decreasing numbers all the way down to where the creek breaks up in the confusion of its delta at the edge of Ledwith Lake. There are abnormal numbers of otter tracks down there, too, at jubilee time; and the banks are whitewashed by unusual gatherings of common egrets there. The activity is clearly all downstream. The animals in the pool obviously come up from that way, and they crowd together in the pool simply because they are unable or unwilling to charge on up through the sluicing culverts.

The essence of the puzzle is, What motivated the migration back wherever it began, and what maintains the motivation throughout the journey? In the Alabama jubilees, oxygen starvation appears to be the factor. Oxygen may be involved in the Wacahoota jubilee too, but this is not easy to prove. As far downstream as the delta there is plenty of oxygen in the creek water. Down where the stream frays out in the marsh and shrub swamp, any systematic plotting of oxygen concentrations is impossible, so sampling has had to be confined to the body of the stream; and this is not where the migration is generated at all, because, as I have said, the migrants are mostly not inhabitants of hammock streams. Many of them are pond animals, typically found in just such habitat as the marshes of Ledwith Lake. They almost certainly come up from there—but why?

Just to comfort myself I have put together a tentative and incomplete theory from the facts at hand. I suggest that the jubilee animals begin their trek because they are overcrowded in the Ledwith pools and that they go upstream because the creek—its distributaries, first, and then the main stream that these unbraid from—are the only avenues for spreading. During times before jubilees the weather was dry. During these dry periods the water level in Ledwith no doubt fell markedly, and sometimes the only water left there was probably that in a few separate bonnet ponds and gator holes, and as these shrank, they lost their connections with each other and with the creek. Perhaps these marsh pools are able to support their teeming refugee biota for a while; but as they keep shrinking, the pressure on their occupants grows, and if the winter rains fail to come in February the fauna will be in serious ecologic trouble from overcrowding. If then a big rain floods the place, there might be a mass tendency to move out into new territory.

It is at that point that my ignorance of the real cause of the jubilees is laid bare, because I am not able to explain how the migration out of the pools into the newly connected distributaries and up the creek is mobilized and maintained. A man finding himself in the fix of the fishes would say, "Let's get to hell out of here. I know of another pond upstream," but pusselguts and sirens are not known to figure that way. Anyway, a rainy spell in late February or March swells the creek with huge volumes of new water, and its distributaries connect up with the isolated ponds; and, for want of a better explanation, it may be permissible to assume that overcrowded marsh-pool animals have a built-in tendency to swim upstream when any sudden affluent breaks into their pond. If so, then maybe they swarm up against the new flow of the tributary creek and just keep going until their way is blocked by the roaring culverts. That is far from being a definitive explanation of the Wacahoota jubilee, because it assumes behavioral traits that the animals involved are not known to have, but it at least collides with none of the known facts.

And sadly, it may have to stand for quite a while, because the Wacahoota jubilees are finished. As they became more widely known in town people started coming out in greater numbers. Their trespassing annoyed the owner and moved him to modify the landscape in ways that changed the character of the flow in the jubilee stream out of Moore's Prairie. I don't know just what he did back there, but since then there has been no Wacahoota jubilee, and apparently there will never be another.

But the jubilee puzzle left its mark on me, and I continue to grope hopefully about for a better explanation of it, still sure that fundamental principles are involved, if only one had the wit to think of them. One approach in my search has been to look about at other exceptional gatherings of different kinds of freshwater animals I could learn of and try to determine whether certain features might be common to them all. So now, every time it rains hard in late winter or early spring I stop at any culvert I can safely pull off of the road by to peer hopefully into the downstream pool. Although nothing approaching the Wacahoota jubilee in complexity and scale has ever come to light, I have gradually realized that Florida fishes of different kinds are actually often stirred to move upstream for reasons that are not apparent. Whether this is the jubilee spirit at work is hard to say, because nowadays the moment that little fishes show up beside any well-traveled road in Florida—and most roads are well traveled—people arrive with nets to scoop them up for bait.

In talking with old residents of the country around Micanopy I found some to whom it is common knowledge that fishes accumulate where culverts discharge the water of storm-swollen streams. It is their opinion that the fishes come to feed. In some cases this is surely true, because some of the gatherings my informants had in mind are not the dense swarms of small species that make up the classic Wacahoota jubilees at all but bands of predaceous sunfishes, grinnels, and soft-shelled turtles that obviously have assembled to grab injured or befuddled small animals tumbled into the pool by the current. Such gatherings used to be common after rains about the mouths of little rills discharging into the Everglades canals. In an old paper of mine on Florida herpetology, under the account of the southeastern soft-shelled turtle, I find this observation: "In the Tamiami canal, they [soft-shelled turtles] may often be seen at the mouths of the tributary ditches, awaiting the occasional objects washed in from the 'glades, and usually in company with gars. Each bit of desirable flotsam provokes violent strife, turtles and gars snapping at each other furiously until the food is torn to pieces and swallowed. It is the chance to exploit a special opportunity that motivates such assembling as that. But the killifishes, crayfishes and leeches of the Wacahoota assemblage were clearly not there to feed."[2]

And yet my talks about the puzzle with the local people dredged up a curiously overlooked memory from my undergraduate days, from a time when I used to drive out to Micanopy and down the Wacahoota Road to shoot ducks on Ledwith Lake. There were Florida ducks and widgeons on the watery plains of pickerelweed, bonnets, and maidencane, and sometimes pintails and green-head mallards, too. Ledwith was controlled by a group of out-of-county hunters. They kept it posted, but Tut Mack was in charge of it, and I had somehow persuaded him to let me in once in a while to pole a little cypress pirogue about in the head-high maidencane and shoot hopefully at the little pods of big ducks that would spring up out of the grass ahead. I still recall how hard it was to get rid of the pole and pick up the gun in time to get in a shot or two. I never killed many ducks, but it was good and lonesome out on the marsh on winter afternoons and the only way I knew to get even close to pintails and Florida mallards. Tut Mack lived just off the dirt road, which was narrower than it is today and ran tunnel-like through the live oak groves and mixed woods of Wacahoota Hammock. And then, as now, it ran by a pool where the road crossed a little creek. Though I remembered the duck hunting all along, what was dragged out of those dim times by my talks with local folk was the faded recollection that one cold evening there were a dozen people—men, women, and children, black and white—standing around the roadside pool ladling out warmouth bass and mud fish with homemade dipnets of screen and chicken wire. I didn't stop. It was nearly dark, and I was no doubt in a hurry to get in to the Presto Restaurant and wheedle Jimmy and John into cooking what was probably the only duck I had shot. I slowly came to recall the look of the people in the twilight bending over the pool and the road strewn with flapping fish that had been thrown there. I don't know why it took so long for that scene to emerge from my memory, because the pool involved, though it was fed by the flow from a single culvert under an ungraded narrow road, was certainly the Wacahoota jubilee pool. And to see the same kind of fishing in progress you only have to go out Florida Highway 315 when long winter rain has raised the level of Orange Lake. You are likely to find the dipnetters shoulder to shoulder along the banks of Orange Creek, where it comes out of the lake and passes under the road on its way to the Oklawaha River. They too are after good-sized game: black bass, warmouths, grinnels, and suckers, and the fish are probably only there to feed. But their being there in unnatural numbers at flood time, at culverts, or at other obstructions of the flow in flooded streams makes me wonder what triggers the upstream travel that takes them to the culvert and whether the jubilee fishes might to some extent, at least, read the same signs. And also they started me wondering whether such gatherings may have been much diminished by the works of man—or, contrarily, whether they might even possibly be produced by some of his disruptions of the landscape.

As always when I want historical insight into a question involving original North Florida landscape, I turn to William Bartram, and this time almost at once I came upon an account of a jubilee. Or rather, I should say, on reading a passage in *Travels Through North and South Carolina, Georgia, East and West Florida* for maybe the tenth time I realized that he was telling of something very like a jubilee.

Of Bartram's encounters with alligators, the most fervidly related are two in which the excitement comes mainly from exceptional abundance of alligators. In both cases there were so many of the animals that Bartram used the same figure of speech to tell of their redundance. He said he could, had he so desired (which he clearly didn't), have easily walked across the body of water involved upon their heads. One of those bodies of water was the St. Johns River at the entrance to Lake Dexter. What Bartram said about the situation there—one of the many stirring things he said—was this:

> How shall I express myself so as to convey an adequate idea . . . to the reader, and at the same time avoid raising suspicion as to my veracity? Should I say, that the river (in this place) from shore to shore, and perhaps near half a mile above and below me, appeared to be some solid bank of fish, of various kinds, pushing through the narrow pass of the St. Johns into the little lake . . . that the alligators were in such incredible numbers, and so close together from shore to shore, that it would have been easy to have walked across on their heads had the animals been harmless? What expressions can sufficiently declare the shocking scene that for some minutes continued, while this mighty army of fish were forcing the pass? During this at-

tempt, thousands, I may say hundreds of thousands, of them were caught and swallowed by the devouring alligators.[3]

This spectacular aggregation of alligators was also baited up by teeming fishes. The conclave, the fish part of it at least, may qualify as a real jubilee. The congregation was produced by the upstream movement of many kinds of fish traveling together, and its late stage occurred in a narrow pass that connected one body of water to another. Although no gathering of alligators ever takes place there today, when Francis Harper was tracing Bartram's route in 1939 he found that the place was still called the Striking Ground because of great numbers of fish that turn up there from time to time.

I had puzzled about jubilees for a long time before it occurred to me that something suggestively similar happens in some of the springs of Florida. The same limestone bedrock in which the sinkholes, solution lakes, and caves of Florida occur gives rise to big rheocrene springs that pour out instant creeks, or even rivers, onto the land. There are about seventeen of these huge river-making springs in the state and hundreds of smaller ones. They were the jewels of the original Florida landscape. One of the regrets of my life is that I was always too poor to buy some of them back in the days when they were still unspoiled and not worth much money. Their like existed nowhere else in the world. Each was a little different from all the rest, with its own combination of animals and garden of plants and its own ways of breaking up light into unexpected colors and of throwing up showers of white sand or of bits of fossil shells.

A few of these springs have been partly saved. One that somehow weathered the generations of dynamiters, speargunners, scuba divers, and churning outboard propellers is Homosassa, the source of the Homosassa River on the Gulf coast of mid-peninsular Florida. Its most distinctive feature is a seemingly fragile and wholly incredible assemblage of different kinds of saltwater fishes that rendezvous in the boil. And this conclave of saltwater fishes must surely qualify as a jubilee.

The first time I saw Homosassa Spring I came suddenly upon it on a winter midafternoon.[4] It was glowing in the dark hammock like a great flat-cut jewel. The light slanted in through breaks among the encircling trees, and the crystal pond shone blue where the sky was reflected, green and maroon over the beds of water plants, and snow white where the current had washed the bottom clean. And throughout the whole great body of the spring were flashes of light glancing from the sides of a myriad of fishes. Barely discernible from my angled view through the tumbling surface over the boil, the fish seemed to be arranged in slowly circling tiers that extended from just beneath the surface far down into the dim depths of the main boil of the spring. I had heard that Homosassa was full of fish, but nothing anyone had told me was preparation for the sight I saw. The point at which I had reached the rim of the basin was close by the main boil, and the water was so roily that I was not able to tell what kinds of fishes were there. Around the shore to my right an old catwalk, held partly up by decaying piling, cut across an edge of the spring. I ran around the shore and climbed out along the tottering boards, hungry for a clear view down into the deep part of the spring. As I crept out on the rotting timbers several streaked-head cooters dropped into the water lettuce lodged among the piling, and a little alligator slid from a partly sunken palm trunk, swam down into a bed of cress and closed his eyes and nostrils as if to hide from me that way. I kept going until I reached a point at which the angle was right and the surface not too roily. I stopped, and I stayed there looking for an hour or more, hardly able to believe what I saw and completely unable to understand it. The whole midsection of the basin seemed filled with fishes—hundreds and hundreds of them, big ones and little ones—of a dozen different kinds. And by far the majority were saltwater species.

    I had looked into Florida springs many times before. I had even seen swarms of fishes in some of them. It had even vaguely entered my head that Florida springs have some strange attraction for fish. But elsewhere it had been mostly local freshwater kinds that were gathered in the spring basins, and in any case it was never altogether clear whether they were actually more abundant in the boils than in other places along the runs, or perhaps just easier to see in the air-clear depths of the head spring. But here at Homosassa there was no such uncertainty. The limpid, tumbling pool was alive with fishes, seemingly almost solidly alive with them; and the main body of the congregation was made up of species that had come up from out in the bay and Gulf. There were orderly swarms of saltwater catfish—both gaff-topsails and sea catfish—pinfish in hundreds, schools of anxious mullet, a dozen big speckled sea trout, and three burly redfish. Down deep in the main conduit I could see mangrove snappers resting, and there were small bands of big sheepshead prying about the rocks. A half-grown tarpon kept charging aimlessly back and forth through schools of smaller fish, which hurriedly opened as he passed and quickly closed behind him; and back near shore under a raft of drift and water lettuce a row of big snooks lay in the gloom like bigmouthed logs. Besides these migrants from the Gulf there were a few big largemouth black bass, multitudes of bluegill bream, little schools of stumpknockers, and a gang of longnose gars. There were other kinds without doubt, but those are the ones I remember seeing from the quaking platform of the rotting catwalk.

I was a fanatic fisherman in those days and was studying ichthyology at the University of Florida besides, and I stayed there on the old walk in a kind of trance. I stayed so long my ribs hurt where the warped boards cut, and I felt dizzy from hard peering at the glancing, blurring shapes, now dim and deep, now flashing silver as they circled in the liquid crystal of the pool. After a long while my angler's instincts took over. I got up, walked the teetering boards to shore, and jogged back through the hammock to where I had parked the car in a clearing. I drove over to Homosassa town, bought a bag of shrimp at the fish dock and some fishing line, hooks, and sinkers at a hardware store, and quickly returned to the spring, where I rigged a hand line with two strong little hooks. I baited these with shrimp, walked the quaking boards to where they passed over the deepest part of the basin, and dropped the bait deep down into the boil. In ordinary water, where you are not able to watch their surreptitious mumbling of your bait, sheepshead are not easy to catch. But in the pellucid depths of Homosassa I could see exactly what the fish were up to and could set the hook by sight instead of by feel. Within ten minutes I had caught five sheepshead—good ones, about three pounds each. As each came up I prudently carried it ashore and hid it under a rotten cypress plank. There was nobody around, and no posted signs, but somebody had obviously been keeping an eye on the place. Otherwise the dynamiters would have wrecked it long before. So I hid my catch as it accumulated, just to be safe. When I had five sheepshead I wound up the line and hid it under the board with the fish, then I went back out the walk and raptly watched the jubilee until the sun got low. Then I picked up the sheepshead went back to the car and drove home in an unending state of wonderment over the marvel I had come upon.

If one thinks about the Homosassa jubilee for a little while, two questions are bound to come to mind. The first is, How can the fishes make the quick change from salt to fresh water without suffering physiological damage? Fishes the world over can be fit loosely into two groups according to their ability to go from salt water into fresh water and vice versa. Some can make the change habitually; some never or hardly ever do. A marine fish has blood that is less salty than the water he lives in and so makes certain physiologic moves that keep him from shriveling up because of water loss through osmosis. For such a fish to pass quickly into fresh water would confront him with a contrasting danger—that of soaking up water into his cells. Salmon, shad, and other fishes that leave the sea and go up rivers to spawn have evolved a special ability to deal with the problem of osmosis. But other fishes, such as jacks and speckled trout, don't make such breeding migrations and seem wholly out of place in an inland spring. There appears to be an answer to this question—one suggested by Howard Odum, a zoologist at the University of Florida. According to him, the dearth of sodium chloride is adequately compensated for by other dissolved solids, particularly calcium salts, in the limestone spring water.

The other question is, Why do fishes congregate in springs? Why do they want to go there? What possible attraction could there be for a sea-fish in a clear spring devoid of food, low in oxygen—surging up out of bedrock caves and huddled about by tall trees? What motivation could there be for such a fish to leave the sea, travel up a springwater river, and consort with crayfish, stinkjims, and stumpknockers away back in the land? Many people have applied their logic to this puzzle, but the solution is not yet in sight. As in the case of the Wacahoota jubilee, explanations that may at first seem plausible dwindle away when you try to make them cover the whole situation and explain the behavior of all the species involved. The most recurrent and most reasonable sounding of these theories is that the fishes go into the springs to "take the waters," as it were—to do curative bathing in them, as our own species has done for thousands of years. Maybe the jubilee fishes have parasites, for example, and the parasites can't stand spring water; so maybe the fishes stop by for a stay in Homosassa Spring to get rid of their vermin, as your grandfather might have summered at Jaybird Springs to get rid of his gout.

The chief objection to that idea—the same one that marred explanations of the Wacahoota jubilee—is that too many different species are involved, with too divergent ways of life, for all of them to have been driven nine miles up a freshwater river to mill about in a spring just to delouse themselves. It would be hard to find three such ecologically divergent fishes as sheepshead, cavally-jack, and gaff-topsail catfish. That these should all have parasites that share a loathing for springwater, that all three should have evolved the instinctive awareness that parasites suffer in springs, and that natural selection has for that reason "taught" their bloodlines that Homosassa is back there in the woods and instilled the guidance instincts to travel there is really too hard to swallow.

Unless, it has only lately occurred to me, the spring-visiting adaptation arose out in the sea itself. Even if visiting Homosassa is actually good for their health fishes would not likely have ever evolved the tendency to take advantage of the benefits so long as the spring was located nine miles up a clear river from the Gulf. But the big springs of Florida are not confined to the land. Many of them boil up out in the sea. I don't know how many submarine springs there are off Florida, but they are out there, and fishes tend to gang up in them just as they do at Homosassa. The big one off St. Augustine has been known for centuries, and I used to spend hours with a water glass looking down into one off Crystal Beach on the Gulf coast.

Before the scuba divers found that one it had a continuous little jubilee in which saltwater fishes of many kinds swam around and around in artesian fresh water piped out from under the land by natural conduits in the limestone bedrock.

So maybe the fishes got the racial habit of visiting springs by barging accidentally into submerged springs in the coastal waters and by then slowly coming to associate some sense of well-being with a stay in springwater—and surviving better for recognizing the correlation. One trouble with that idea is that it leaves unidentified the specific cause of the sense of well-being. Another is that it says nothing about the pathfinding problem involved when the spring lies far back inland.

In searching for an explanation of the Homosassa jubilee it seems reasonable to look at other Florida springs to see what similarities and differences there may be in their fishes, and in their connections with the sea. Besides mullet, tarpon, and snook—which are all strong swimmers and well known for their ability to pass back and forth between fresh and salt water—two other marine species, both of them small and sedentary, are found in most big springs. One is a pipefish; the other is a sole. Both occur in practically every spring run in north-central Florida, in both the Atlantic and Gulf drainage. Nevertheless, it seems doubtful that they really qualify as jubilee fishes. Spring-fed streams are simply part of their natural habitat. The pipefish likes the beds of submerged thin-leafed water plants, and the sole is made happy by the patches of sand washed there by the current. Both of these little fishes have relatives that occur in fresh water in other parts of the world. In most of the springs tributary to the St. Johns river there are stingarees and blue crabs, but because they are regular sojourners in springs their presence there may have nothing to do with the jubilee phenomenon.

The east coast springs fit into the jubilee picture in a somewhat different way. None of their runs goes directly into the Atlantic Ocean as that of Homosassa goes into the Gulf. Instead, they all enter the St. Johns River. Faunally speaking, the St. Johns River is an extraordinary stream, like no other in America.[5] Its most distinctive feature is the degree to which marine animals of diverse kinds and divergent ways of life either live there permanently or seasonally ascend the river. One can readily see why shad and herrings go up the St. Johns and why bass and shell-crackers and red-bellied cooters are there; but what are the croakers and redfish doing in the croaker hole? Why do the beach-hoppers skip about the shores of Lake George 115 miles above the mouth of the river? The St. Johns, above Jacksonville, where the salinity has fallen to less than one part per thousand, has 170 species of fish in it. Fifty-five of these are freshwater fishes that obviously belong there. Of the rest many are anadromous kinds—that is, kinds that habitually move up rivers to spawn, as the shad does. But besides these there are many others whose business in the St. Johns is less clear, or not clear at all. Some of these are euryhaline kinds, those that are able to pass freely between salt and fresh water; others are not noted for this ability. Some of both kinds breed within the river itself; others go back to the sea to reproduce. In the aggregate, the St. Johns fishes make the stream one of the most interesting rivers of the world.

The St. Johns begins far down the Florida peninsula on the Kissimmee Prairie, only fifteen miles from the Atlantic Coast, and runs northward, never far from the coast, for 260 miles to its mouth above Jacksonville. It falls only twenty feet along its course, and though it is the master stream for a vast area of the peninsula, it bears little silt. The water is stained by tannin but is relatively clear. Here and there the river spreads into lakes and lagoonlike reaches, the biggest of which, Lake George, is fourteen miles long and four miles across. One of the peculiar features of the St. Johns is a double salinity gradient. From the mouth up to Palatka the salt content falls regularly. From there on, however, through a section of the river that receives the discharge of a number of mineral springs, salinity begins to rise again. The increase continues until Lake George is reached. Above Lake George the salinity falls again.

The peculiar fauna of the St. Johns River has apparently been produced by two kinds of factors: ecologic and historical. A part of the extraordinary penetration by marine species can be attributed to the dissolved salts from the springs, which help keep the saltwater fishes out of osmotic trouble. Besides offering ecological conditions under which saltwater fish find it easy to survive, the river has evidently had a past in which traffic between it and the sea was easy and continuous. Sections of the St. Johns are apparently old coastal lagoon. Long reaches of the river almost surely began as estuaries like the Indian River, Halifax River, and Mosquito Lagoon of today. As the sea level fell or the land rose—or as both occurred—the old St. Johns lagoons moved inland and became the master stream in the drainage of the country east of the Central Ridge of the peninsula. Normally this new role would have caused the river to freshen rapidly upstream. But, because heavily mineralized water comes into its middle reaches from the springs, the midsection stayed hospitable for a lot of fishes that would otherwise have moved up only a short way above the mouth.

Thinking of the ecology and paleogeography of the St. Johns together like that makes the croakers in the croaker-hole seem less bizarre and tells why there are commercial blue-crab fisheries in Lake George. It also may help explain the small jubilees that form in the St. Johns River springs.

Perhaps the habit that these fishes formed was not that of ascending a river for a hundred miles to visit a spring but simply that of congregating in that spring when the river was a coastal lagoon. And possibly the tendency was just held over from an even earlier time, when the St. Johns springs were under the sea.

Maybe the Homosassa jubilee formed the same way: as a submarine spring that slowly migrated inland and easily took its jubilee along with it because succeeding generations of the jubilee fishes had only to adjust to a very gradual lengthening of the run.

This still leaves unexplained the fundamental questions: Why do sea fishes rendezvous in freshwater springs, both out in the ocean and back inside the land? Why do freshwater fishes make other pointless-seeming, up-current treks? Why do myriad pond fishes of a dozen different kinds join crayfish, leeches, and salamanders in mass migrations up a hammock stream?

Whether there is real kinship among these unexplained migrations cannot be told. But in any case, they are all beguiling mysteries of natural history. Though I gladly acknowledge Daphne's prior claim to the term, I shall think of them all as jubilees.

Sticky Heels

The first box turtle I ever saw was eating a cactus fruit on the plains of central Texas. I was six years old and had known tortoises only from some sketches in a child's book about animals. I well recall my astonishment when the turtle stretched its neck to blink myopically at me with the same mild, meditative look caricatured in my animal book.

I dashed forward to seize the exciting creature before it should escape, but too late—for what I picked up was only an artifact, sculptured with skill and gaudily rayed and flecked with color but surely inanimate as chert. Sensing that mistrust of me had brought the metamorphosis, I put the thing down and waited minutes, and at last the shell opened ever so little to show a pair of orange-spotted wrists and the tentative glitter of an eye in the depths between them.

I shifted my position slightly to get a better look, and the shell shut abruptly. Another wait, with more restraint this time, and with the final reward of a wider gap in the portal. Stealthily, I advanced a twig from a blind side and by luck managed to slip it between the upper and lower shells in front. Triumphantly I picked up the vanquished tortoise and, substituting for the stick my own fingertips, tried to widen the breach by a sudden pull. This was a mistake. The shell snapped shut on my tender digits, and my stock of box turtle lore was multiplied in an instant.

My father, who was hunting quail a quarter of a mile away, ran the whole distance, convinced that my howls of anguish could only mean I had stumbled over a rattlesnake. By the time he arrived I had freed my fingers from the vise that had held them, but they kept on hurting for a long time to incise a parable that to this day has sustained in me an extra measure of respect for the privacy of others.

In a footnote to the manuscript of this chapter Archie states, "Sticky Heels is what the Delaware Indians called the box turtle because of the slow manner in which he lifts his hind feet in walking." I don't know where he found this information, but I trust him.

I have recently been engaged in writing a handbook of turtles.[1] Such a book is of course largely a compilation of the work of others, and perhaps the most trying part of the job is searching and making notes on the pertinent literature. Like many people, I make notes on three-by-five cards, recording each reference to a given turtle on a separate card. In my file, the box turtle group (the genus *Terrapene*) is represented by three inches of cards, and the common box turtle alone has one and five-eighths inches. No other turtle has such representation except the common snapper (with two and a quarter inches), but this gross beast has the combined advantages of an arresting aspect, a vile temper, and extreme succulence of flesh, while the box turtle must stand on its personality alone.

Although this device for measuring our knowledge of the box turtle is crude, it has a certain validity. Everybody likes box turtles, and we've lived with them for several centuries now. Dozens of naturalists, both lay and professional, have studied and recorded their ways, and there aren't ten reptiles in the world for which a more respectable natural history could be pieced together. Nevertheless, in the résumé that I give below the gaps are astonishingly conspicuous, and I hope they will point up the moral that anyone with a backyard and a box turtle can still make a real contribution to zoology. The notes that follow refer to the eastern race, *Terrapene carolina carolina*.

HIBERNATION. In the northern part of its territory the box turtle hibernates in leaf mold, loose soil or sand, or even in the mud of stream bottoms during the severest part of the winter. Published records outline the season as extending from October to April, but there is so much local variation that dates have limited general significance. Little is known of the physical factors that induce hibernation and emergence. Individuals may burrow progressively deeper as winter deepens, and they may emerge temporarily in spells of unseasonably warm weather.

AESTIVATION. No protracted periods of summer torpor have been observed, but individuals often burrow to escape drought or the midday heat of summer. The danger of drying up is a real one to the box turtle, especially to the young, which are quickly immobilized by desiccation. Nothing is known of the limits of temperature tolerance or of the normal body temperatures of the active box turtle.

AQUATIC TENDENCIES. I have already said that hibernation sometimes occurs under water. A temporary assumption of aquatic inclinations appears to substitute for a period of aestivation in many cases. During such periods the turtles loaf about in the vicinity of small streams, occasionally entering them to swim awkwardly about or even to walk around on the bottom.

HOME RANGE. Box turtles are homebodies, with no touch of the gypsy in them. They often spend years within an area a few hundred yards square, and if removed from home territory by the hand of man to points as much as half a mile or more away they often find their way back with little delay.

FEEDING. The box turtle is omnivorous. As in many other turtles, however, growth from hatchling to maturity brings a progressive change from a mainly carnivorous diet to one in which plant food predominates. Another noticeable feeding tendency is a sort of opportunistic exploitation of whatever food happens to be abundant at a particular time. Box turtles have been known to stuff themselves on berries or mushrooms until they are unable to close their shells. It appears to be a fact that they eat the ghastly Death Angel mushroom *(Amanita phalloides)* with impunity, and it has been suggested that this habit may explain the poisonous qualities occasionally attributed to box turtle meat by those who have eaten it.

ENEMIES AND ENVIRONMENTAL HAZARDS. Although box turtles have the most complete bony armor of any vertebrate animal, even it does not insure the breed against violence from without. As in most cases, people are the most inexorable enemy, and the number of box turtles smashed each year on the highways is imposing. Although it is probable that few natural predators succeed in cracking the strong shell of the adult, the hatchlings would seem to be vulnerable to whatever carnivore might succeed in locating them, and it is probably this danger, combined with that of desiccation, that makes young *Terrapenes* so amazingly secretive. The nests are robbed of eggs by a long list of animals. The shells of box turtles are sometimes found in the woods in large numbers, and two explanations for their occurrence have been proposed. One suggests that they are the remains of individuals caught by the sudden return of stupefying temperatures after periods of mild weather have caused them to emerge from hibernation. The other attributes the mortality to males' inability to right themselves in soft leaf mold after they fall over backward in the customary denouement of copulation. Ticks, mites, and various internal parasites plague box turtles, and the grubs of dipterous insects often infest them. A fate that impresses me as extremely bizarre comes to the *Terrapene* that loafs too long on the mud of a disappearing pool. The turtle may stick at last in the plastic clay and then day after day kick with pitiful incessancy to excavate a moat surrounding a clay pedestal, on the pinnacle of which it ultimately expires.

Box turtles hold a strange fascination for dogs, especially hunting dogs. An otherwise flawless pointer—rabbit-proof and staunch as a rock—will often freeze in a rigid point at the near scent of a *Terrapene* and then droop in bewildered humiliation when his error is shown to him. The only other creature I have seen deceive a bird dog so utterly is a little ground-dwelling bird (presumably a sparrow, though I never have had a good look at one) that is so often mistaken for quail that it is known throughout the southern states as stink-bird. Hounds, too, are demoralized by box turtles, but in a different way. Whether because of the scent or the challenge of the unassailable shell, hounds often drop whatever business is at hand to attend to a box turtle. One dog may stop and lie quietly, muzzle between paws, to regard a disinterested tortoise with unaccountable emotion, or even raise his head from time to time to voice his feeling in the low, soft, eerie moan of a bereaved oboe. Another may bury the turtle at once, and another simply carry it about for the rest of the day. I have several times known the depths of frustration when on a deer hunt a star hound left the drive to trot brightly up and show me a box turtle that he had found, confident of applause and pained when it failed. Occasionally a dog tries his hand at gnawing into the shell, but I have no recollection of seeing a successful assault of this kind unless the unfortunate turtle had been cracked on the highway to begin with.

DISPOSITION AND MENTAL ATTRIBUTES. Box turtles are generally mild and inoffensive in their attitude toward humans, but once in a while one may show an inclination to bite. Among themselves they exhibit all shades of disposition. The study of box turtles in captivity holds no more stimulating promise of reward than that to be gained from appraisal of their psychological traits. Enough data have accumulated to show that *Terrapene*, though no mental giant, is capable of elementary departures from its ingrained patterns of response. It has memory and the ability to learn and make choices and to form relatively complex social structures when grouped with its fellows into unnatural communities. Cleverly devised tests and systematic programs of observation of captive individuals and communities need not be regarded as the prerogative of the trained psychologist; they can be made by anyone, and with the assurance that he is unearthing data of scientific interest, as well as entertaining himself.

COURTSHIP. On emerging from hibernation the male is usually ready to begin a straightforward kind of courtship, during which he walks about behind the female, often only a short distance away but sometimes so far back that it is thought that he may follow by scent. When the female loses some of her coyness and allows the male to approach he bites at her shell, head, and neck and sometimes noses at the lower part of her shell, as if trying to upset her. One observer reported that the male may tap the shell of his prospective mate with his own carapace.

COPULATION. When the female has become acquiescent the male mounts her shell from behind and hooks his hind feet under the hind edge of her carapace, where her own hind legs usually hook around them. Actual coupling takes place only when the male elevates the fore part of his shell until he has assumed an almost vertical position, and the act often terminates with his falling over backward.

DEFERRED FERTILIZATION. The normal period elapsing between fertilization and deposition of the fertilized eggs is unknown, but female box turtles, like other species, may store spermatozoa after copulation and lay fertile eggs up to four years later with no further attentions from the other sex.

NESTING SEASON. The females begin looking for nesting sites in May, June, or July, the exact time varying with latitude and also, no doubt, with individual eccentricity. Females may prospect for some time before selecting a location. Actual excavation of the cavity usually begins in the late afternoon or early evening and may be continued until late at night.

THE NESTS. When soil of the proper texture and consistency has been located (and there is much individual variation among the females as to their notions of what constitutes a proper nesting soil—everything from lawn sod and rotten wood to almost sterile sand have been used by individuals under observation) the laborious operation of digging the cavity begins. The process is an exercise in deliberate motion that is ludicrous in the extreme to watch. The stubby hind feet work alternately, each in turn scraping at the sides and bottom of the imperceptibly growing pit and then stopping to remove the loosened earth by grasping a few grains at a time, lifting them out and dropping them at the sides or rear edge of the hole. The process may require hours. The resulting nest is roughly flask-shaped and of a depth about equal to the distance to which the hind foot can be extended. The eggs are dropped into the hole at intervals of one to six minutes, and each is arranged in the bottom as it falls. Covering and filling is accomplished by raking movements of the hind feet, which also tread and trample the soil to pack it in.

THE EGGS. The egg complement varies in number from two to seven and is usually four or five. The eggs are elliptical and are on the average about one and a quarter inches long by three-quarters of an inch wide, with white brittle shells. It is not known whether the female normally lays more than once during a season or if a correlation exists between the number of clutches laid and the size of the individual clutch.

INCUBATION. The length of the natural incubation period is unknown. In captive specimens the shortest period on record is seventy days. Data on the opposite extreme are obscured by the fact that emergence from the nest, and probably hatching, are sometimes deferred until the spring following oviposition. Slight changes in temperature and humidity almost certainly produce conspicuous variations in the embryologic schedule. This whole question is in need of further investigation.

THE YOUNG. The hatchling box turtle is usually about one and a quarter inches long and one and one-tenth inches wide, with a low, keeled shell marked by a comparatively small number of light spots. For some time following its emergence from the egg the tip of the snout is adorned by a thornlike egg tooth. Considerable mystery surrounds the activities of the hatchlings, and they are almost never found by collectors.

GROWTH, AGE, AND LONGEVITY. The popular faith in the longevity of chelonians receives some support from what is known of *Terrapene*. Although growth is relatively rapid at first—captive individuals having gained as much as three-quarters of an inch in shell length in one year—it slows down progressively after sexual maturity is reached during the forth or fifth year of life. J. T. Nichols, who probably knows more about the subject than any one else, estimates the normal life expectancy to be between forty and fifty years and believes in the existence of occasional centenarian box turtles.[2]

The oldest *looking* turtle I ever saw was a *Terrapene* that I found beside a highway across Kissimmee Prairie in south-central Florida. As I drove along with an eye alert for snakes I had a fleeting impression that a lump on the road shoulder had moved. I stopped the car and walked back, and I found a box turtle to marvel at. Even with the creature in my hand I had to recall that I had seen it walking to believe that it was alive. It was a twisted dwarf of a turtle that seemed to have met with some fearful, stunting catastrophe ages before and to have grown since only in grotesqueness.

Its shell might once have been melted and thrown at a wall while cooling; it was the shell of a clay turtle smashed by the fist of a disgruntled modeler, now bulbous and eccentric, with the right half bulging into the left and all gnarled and craggy, like growing coral or weathered driftwood. Healed but still traceable, a long cleft marked the former median line of the carapace and intercepted another deep crack that crossed the sunken right side of the shell like an ancient overthrust fault. The horny laminae and their pigmented pattern were missing, except here and there, where patches of scale substance were daubed and streaked on the staghorn surface like thick varnish from a too-stiff brush.

The undershell was partly gone, its hind piece rigidly fused at the hinge; the fore section was still movable but unable to close tightly since it no longer fit the warped upper shell. About the fore and hind legs of the right side there was no shell at all, and on these the feet too were missing.

How many years and how much disaster had molded this pitiful figure of a tortoise? He had been crushed, but was it by the iron-shod wheel of a covered wagon or by the rubber tire of a fish truck? He had been grievously burned, but was the dry prairie kindled by the Seminoles fleeing Jackson's men or by the ranchers who use it today?

As I held the formless object in my hand and meditated on it, half ready to admit that I had never seen it move and never would, the front lobe of the plastron lowered slightly. This was an unmistakable symptom of curiosity, meaning not merely that the creature was alive but that he fermented with an extroversion uncommon in his kind. I set him aside and watched. Shortly, a scarred beak appeared, and then the head slowly emerged and rose as the long neck stretched, and my gaze was met by that of a pair of clear amber eyes, bright and unperturbed and unclouded by any memory of past tragedy or fear of tragedy to come.

When the tortoise had seen enough of me he turned his head toward the palm-set horizon and blinked a while and then began to walk, good legs alternating with footless stubs to trundle the incredible body along in a fittingly lopsided gait. A turtle crudely carved from a lightwood knot and perversely come animate, he plodded out across the prairie, impelled toward who knows what goal and destiny by hope from what hidden spring.

A Florida Picnic

During 1980 and 1981 we operated a very successful buzzard feeder out at our farm. It was a rewarding enterprise and one that convinced us that more goes on in the crocodilian mind than was dreamed of in our philosophy. It was none of the humans in our family who brought about our carrion bird-feeder. It was our dog, Ben. Ben was a German shepherd who weighed one hundred pounds and had canine teeth that once were long but were eventually worn down to stumps by repeated gnashing at the plated backs of armadillos. Ben reached his prime during the period when armadillos were at the peak of their unnatural abundance in our woods, and he spent most of his time chasing them through the forest or trying to dig them out of their holes. He even woke us up nights, whining to get out to punish armadillos he heard or smelled or clairvoyantly sensed the presence of in the flower bed. His concern with armadillos was obsessive.

Though Ben rarely could catch an armadillo in the woods, if one crossed the yard or pasture he usually ran it down. The gape of his jaws was vast, and when his teeth were sharp he could easily bite through the armadillos' armor. So on many mornings the corpses of these poor dead armadillos were lying around, usually out in the calf pasture—a quarter-acre field of mown Bahia grass behind the house—where they had lost the race with Ben. The vultures soon discovered the bonanza and began coming in every day to check out the place. They couldn't do much with a corpse until a little decay loosened its armor, so they perched about on limbs of trees at the pasture edge and waited philosophically until the time—or, rather, the prey—was ripe for a picnic. Thus it was Ben and the buzzards themselves who thought up our buzzard feeder.

Every morning we would go out and look for dead armadillos, and if we found one we moved it to the middle of the pasture. Usually by the next day the buzzards would be at work—a couple dozen of them, or sometimes

From *Animal Kingdom* 85, no. 6 (1983).

even more—by turns pecking and pulling at the chinks in the armor of the corpses. There rarely was enough armadillo to go around, but there were no serious disputes among the vultures. There was some shouldering aside or a short chase now and then, but mostly it was a "now it's my turn to try" sort of give-and-take, and this was an agreeable thing to see.

At first the most abundant visitors were the black vultures, but soon a few turkey vultures started coming. When one of them would glide down, survey the scene briefly, and hop over to join the group, the black vultures would give way to their more assertive relative without protest. They would stand around waiting until the turkey vulture ate all it wanted or got tired of trying to get inside the corpse.

So our vulture feeder was very rewarding and not at all offensive because it was located northwest of our house—a direction from which the warm breezes of the armadillo season never came.

A trio of fish crows that frequent our farm used to come down once in a while and, by virtue of their spryness, filch morsels from under the buzzards' noses; and to add to the action out there, Ben would gallop out, roaring, in an occasional fit of petulance, scattering the picnickers. We could tell he did it partly because he felt the armadillos were his and partly just to see the buzzards scatter. These attacks were no great worry for the birds. They just flapped up into nearby trees, and when Ben stopped bellowing they swooped back down and resumed feeding.

Then one day there was a strange shape at the carcass. It was bigger than the vultures and browner, and it clearly outranked them all in the pecking order, because they stood around in a ring disconsolately watching it eat. We got the binoculars and saw that the newcomer was a bald eagle—a young one not yet in its black and white plumage but with a commanding presence all the same. It looked wonderfully stern and raptorial among the buzzards, and it was obviously delighted with the armadillo.

That put us into a frenzy of scrabbling for dead armadillos—begging our friend who gets up at night to shoot them in his garden to stay up longer; risking our necks to stop at corpses on I-75; making Ben sleep outside, though he loved the air conditioning. We began getting more armadillos, and it paid off. Nearly any time we could put two or three corpses out on the grass, an eagle came. The eagles really seemed to have an inborn predilection for armadillo. The peak of our success was looking out one noon and seeing, in the ring of grieving buzzards, three eagles—one brown young and two adults in resplendent black and white regalia.

That was the peak of our success, but not the climax of the story. The climax came one day when we looked out and saw, in the circle of vultures, not an eagle but a nine-foot alligator. It was our alligator; through the years, that alligator had done many things that prove the versatility and wisdom of crocodilians, but I think finding the buzzard feeder was the most astonishing.

I found her trail through the maidencane grass where she had come out of the pond, and I paced off the distance up through the woods to the pasture. The pond is located down a slope from the carrion bird-feeder, and its surface is at least twenty feet lower than the pasture. So all you can see from alligator-eye-level is woods sloping up to an unseen field ninety yards away. But she found the feast anyway. She seized a carcass in her jaws, shook it violently around—to the consternation of the buzzards—and then walked high-bellied and unhurried down through the woods to the exact place where she had come out of the pond.

After that day, the old alligator did the same thing repeatedly, so many times, in fact, that we finally had to give up our feeder. Nights got cold and armadillos got few, and nearly every time we were able to put one out in the field the alligator got it. This not only made it hard to keep carrion out there, it also compromised our ethics. We are a family that is dead against feeding alligators—above all, ones you have to live near. We subscribe to the stern belief that it is better for everybody, the alligator included, if you throw a rock at it instead of a hamburger bun. We had never fed ours a bite of anything—except inadvertently, when she broke down a pen and ate our flock of mallards. Feeling that way and having widely preached the doctrine, we had to quit providing our alligator with armadillos.

Now, I am not seriously saying that our alligator, or any other crocodilian, is really wise, but her sensory—or intellectual—feat of plotting a direct course through heavy woods from the low pond edge to a high pasture, where unseen armadillos lay, is not an easy thing to explain. A smell trail might seem the most conceivable source of her guidance; but if you had ever watched the smoke from our garbage pit behind the house swirl and drift around the place, you would know there was no organized smell gradient leading from the corpse down through the woods, along the course the gator followed.

The only way visual cues might have helped was by her having seen buzzards and eagles circle and land. The only way the sense of hearing could have operated was by hearing us yell that eagles were at the feeder.

So, with this unsettling evidence that crocodilians think thoughts we know nothing about, I naturally was excited when my friend Larry Ogren reminded me of the sapience of the mugger of Mugger Ghat, the immense crocodile in Kipling's story.[1] The mugger had grown old and attained a length of twenty-four feet eating the people of an Indian village, and he explained his skill in catching them—"my people," he termed them—in these words: "Is there a green branch and an iron ring hanging over a doorway? The old mugger knows that a boy has been born in that house, and must some day come down to the ghat to play. Is a maiden to be married? The old mugger knows, for he sees the men carry gifts back and forth; and she, too, comes down to the ghat to bathe before the wedding, and—he is there."

The old mugger saying those things was fantasy, of course, but the things he says show plainly that Kipling was a close observer of the ways of crocodiles.

All the Way Down upon the Suwannee River

The name *Suwannee* was spread so widely by Stephen Foster's song "Old Folks at Home" and the generations of black minstrelries it inspired that one loses sight of the fact that outside of northern Florida and southern Georgia the river is not very well known. Back when Florida was being discovered, explored, fought over, made into a state, and settled, it was the St. Johns that provided the chief access to the interior. The Suwannee was not a convenient highroad to anywhere. Its mouth was way down in a remote corner of the Gulf of Mexico, out of the main ship route. The delta region offered no real harbor, and there was little high ground there, so no port town developed. To this day the whole Suwannee Valley is a rural region of farms and small towns just beginning to feel pressure from the development rampant in the rest of the peninsula.

The most far-reaching fame of the river comes not from its many intrinsic merits nor its very lively history, but from the fact that *Suwannee* is a trisyllable that can be slurred into a disyllable which, when stressed on the first syllable, fit the meter of a song that Foster was trying to write. The fame of the song spread the fame of the river.

Some say the name Suwannee arose as a corrupted form of the Spanish San Juanito, or San Juanillo. On the other hand, there is a north Georgia town called Suwanee, and the Cherokees who named it probably got the word from the Creeks, in whose language *suwani* means echo. In any case, before the name Suwannee was widely accepted the river had been given a lot of other names: San Juan, Little San Juan, San Juanito, Guacara, San Juan de Guacara, San Pedro, San Pierre, and San Martin. If any of those had stuck, Foster would have looked elsewhere for his river, and the history of the Suwannee would have been different somehow. And whether Foster ever saw the Suwannee appears to be moot. Some say he visited Ellaville back when it was a booming town, but that may be wishful thinking.

Adapted from *Audubon* 85, no. 2 (1983). Copyright © 1983 by the National Audubon Society, 700 Broadway, New York, N.Y. 10003.

When you see a road sign announcing the nearness of the "Historic Suwannee River," what the sign painter had mainly in mind was that the Suwannee looms large in the folk music of the land. But actually the river is historical in its own right. Before the Spaniards spread missions and ranches about, the Suwannee was the border between territories of the Timucua and Apalachee Indian confederations. Narvaez crossed the river in 1528; and in 1539 de Soto fought a big battle at the Indian town of Mapitaca, near Live Oak. Using lances and crossbows, his mounted horsemen killed forty Indians and captured three hundred. Down in the Suwannee delta, pirates lurked in the maze of waterways. There were Spanish settlements and missions all along the river and its tributaries. At least one penetration by the French is recorded. In 1682 Don Tómas Menendez Marquez was captured by French buccaneers at his Hacienda La Chua (Alachua), far up on the Santa Fe River. He was held for a ransom of one hundred fifty cattle. The Seminole Indians came into being when Secoffee led a band of dissident Creeks out of Georgia to settle along the Suwannee and in adjacent parts of the fertile interior lands called Alachua by earlier Indians. Long before traffic between peninsular Florida and the American colonies had been established, Suwannee River Indians were making ocean voyages in dugout canoes. This is William Bartram's account of that traffic, recorded in 1773, during his stay at the Indian town of Talahasochte, on the banks of the Suwannee near Manatee Springs.

> These Indians have large handsome canoes, which they form out of the trunks of Cypress trees *(Cupressus disticha),* some of them commodious enough to accommodate twenty or thirty warriors. In these large canoes they descend the river on trading and hunting expeditions to the sea coast, neighboring islands and keys, quite to the point of Florida, and sometimes cross the Gulf, extending their navigations to the Bahama islands and even to Cuba: a crew of the adventurers had just arrived, having returned from Cuba but a few days before our arrival, with a cargo of spirituous liquors, Coffee, Sugar, and Tobacco. One of them politely presented me with a choice piece of Tobacco, which he told me he had received from the governor of Cuba.
>
> They deal in the way of barter, carrying with them deerskins, furs, dry fish, bees-wax, honey, bear's oil, and some other articles. They say the Spaniards receive them very friendlily, and treat them with the best spirituous liquors.
>
> The Spaniards of Cuba likewise trade here or at St. Mark's, and other sea ports on the west coast of the isthmus, in small sloops; par-

ticularly at the bay of Calos, where are excellent fishing banks and grounds; not far from which is a considerable town of the Seminole, where they take great quantities of fish, which they salt and cure on shore, and barter with the Indians and traders for skins, furs, &c. and return with their cargoes to Cuba.

The trader of the town of Talahasochte informed me, that he had, when trading in that town, large supplies of goods from these Spanish trading vessels, suitable for that trade, and some very essential articles, on more advantageous terms than he could purchase at Indian stores either in Georgia or St. Augustine.[1]

Much of the dismal aftermath of the War of 1812 and the action of the First Seminole War took place along the Suwannee. In 1817, in a lawless move to recover slaves who had escaped from owners in Georgia and South Carolina, Andrew Jackson attacked and defeated a combined force of slaves and Indian allies at Suwannee Old Town. In April 1818 he attacked Suwannee, Boleck's (Billy Bowlegs') town on the upper river above White Springs. When he arrived he found that Boleck had fled into the vastness of the Okefenokee Swamp, having been warned of the Americans' coming by the son of Arbuthnot, a Scottish friend of the Indians. As a result of this incident and Jackson's generally atrocious temper, the elder Arbuthnot was tried and hanged.

During the Civil War the most important action in Florida was the Battle of Olustee, fought when five thousand federal troops, on their way to destroy a railroad bridge, were met by twenty-five hundred Confederates under General Joseph Finegan. The battle lasted two hours, and the Confederates were victorious. Later, the demise of the Confederacy took place on the Suwannee River. On May 15, 1865, a curious cavalcade moved into northeastern Florida. It consisted of a heavy wagon and an ambulance, both mule-drawn, guarded by nine military and civil officers of the Confederate States of America—"members of the most distinguished families of Maryland and Louisiana." Two scouts rode in advance, and five black servants brought up the rear. The train bore the last material remains of the Confederate States of America—the government archives, the baggage of the accompanying officials, and thirty-five thousand dollars in gold. The Confederate government had dissolved a short while before, at Charlotte, North Carolina. Hope of its revival lingered in Texas, however, and it was toward the west that the train was traveling; the Florida route was chosen because the Suwannee Valley was mainly wilderness. The long, dangerous journey down from North Carolina came to an end on May 22 at the Yulee Plantation on the Suwannee near Madison, Florida. There, the last followers of the Confederacy learned that President Davis had been captured and that further travel toward Texas was pointless.

During the early 1800s steamboat traffic developed on the Suwannee. Paddle wheelers were used in the endless pursuit of the Indians in the two Seminole wars. When Fort Fannin was built there was regular weekly steamboat service between Fannin Springs and Cedar Key, ten miles south of the mouth of the river. During the Civil War, riverboats were important in supplying and deploying federal troops and were used as Confederate blockade runners. After the war they carried cedar logs from the Suwannee country down to the Cedar Key pencil factory.

For half a century the river boats of both the Suwannee and the St. Johns were more important to transportation in Florida than all the desultory roads and slowly growing railroads combined. The most colorful of the Suwannee steamboats was the *Madison,* owned and operated by the redoubtable Captain James M. Tucker. The *Madison* made trips once a week from Cedar Key upriver as far as water levels would allow. Usually it stopped at Grab, two miles above Troy Springs, but sometimes it went on up as far as Columbus, a town that used to be across the river from where Ellaville is now.

The part the *Madison* played in the life of its time is engagingly sketched in an old newspaper article by John M. Caldwell, publisher of the *Jasper News,* who knew the vessel well. According to Caldwell:

> The *Madison* carried a line of general merchandise which was traded to the settlers for money, venison, hams, cow hides, deer skins, tallow, beeswax, honey, chickens, eggs, hogs and beeves. There was no warehouse on the river and the boat would tie up at a landing and stay as long as the people wanted to trade and then move on to the next landing. The *Madison* had a whistle that could be heard ten miles and this whistle was blown at intervals to give the people time to reach the landing with their produce.
>
> When the Confederate war began Captain Tucker raised a Company of the Confederate soldiers and took them aboard the boat. This company slipped out of the river one night on the Madison and captured a Federal gunboat. Finally orders came for Captain Tucker and his company to go to Virginia where the company was afterwards known as Company H, 8th Florida Infantry.

Caldwell says that the *Madison* once traveled as far upriver as White Springs. Captain Tucker badly wanted the Suwannee to be declared navigable up to White Springs; but efforts to bring this about had failed because no boat had ever traveled that far. When Captain Tucker heard this, "he swore he'd be damned if he didn't put the *Madison* in White Springs if he had to run her on wheels." One day "it began to rain . . . and [it] kept on raining, and the river kept on rising till the Suwannee overflowed its banks and ran away out in the woods." Captain Tucker got up steam in the *Madison* and set out for White Springs. He got there, and he got back. His smokestacks and pilothouse were raked off by the trees, but by the time the *Madison* was repaired the Suwannee River had been declared a navigable stream from its mouth to White Springs.

In 1863 the *Madison* was scuttled in Troy Springs so she would be hidden until the Civil War was over. But before the conflict ended, her boilers were carried away to the Cedar Key salt factory, her funnels were cut up to be used in sugar furnaces, and her cabins were dismantled for their lumber. You can still see the skeleton of the *Madison* on the bottom, beneath the crystal water of the spring.

The Suwannee River arises in the Okefenokee Swamp, or Oquaphenogaw Swamp, as Bartram chose to call it. The Okefenokee is 675 square miles of shallow lakes, wet prairies, cypress swamps, pine islands, and trembling earth. The swamp lies mostly in Georgia, but its southeastern margin edges over into Florida, where it is continuous with Pinhook Swamp. Drainage eastward to the Atlantic is partially blocked by the Trail Ridge, a relict barrier island; and the main outflow from the Okefenokee is by way of the Suwannee River, which emerges on the western edge and flows southwestward to the Gulf of Mexico. At times water from the swamp drains into the headwaters of the St. Marys River, which flows into the Atlantic near Fernandina Beach. Within the Okefenokee Basin, separation of the headwaters of the Suwannee and the St. Marys may sometimes be no more than the highest hump of wetness from the last rains. I doubt that anybody has ever done it, and I would not want to encourage it, but after very heavy rainfall you ought to be able to enter the mouth of the St. Marys and ascend it in a canoe, paddle and claw your way around the lower end of Trail Ridge and out into the swamp to the gathering place of the Suwannee River water, go from there down into Narrows of the River, portage over the Sill, where the Suwannee leaves the swamp, and then, after paddling for a week or so, arrive at Suwannee Sound and the Gulf of Mexico. That would be a desperate journey, but the chief obstacle would be getting lost among the mazes of water lanes and gator trails of the swamp, not a lack of water to travel on.

The Okefenokee is a national wildlife refuge. Though it has been much modified by the timbering that began more than a hundred years ago and took away most of the virgin cypress and pine, it is still a unique, lonely, and lovely spread of wet wilderness. The wildlife is diverse and includes most of the vertebrate animals of the southeastern coastal plain, a region noted for superior animals. On sunny days in early spring you can see more big alligators in Billy's Lake in the Stephen Foster State Park than anywhere I know of. When they come out of their winter lethargy and have not yet dispersed to breed, they mix with teeming boats full of Georgia people fishing for redbreast bream in a peaceable kingdom that is a triumph of laissez-faire in park management. During much of the first half of this century botanists and zoologists from Cornell University made repeated exploratory expeditions into the Okefenokee, and an impressive literature on the natural history of the swamp grew out of their findings.

One of the best short stretches of canoe water on the whole Suwannee comes just before the river leaves the swamp. This is Narrows of the River at the end of Billy's Lake, where the gathering stream heads down toward the Sill. Here there is an enchanted half-mile where you travel a twisting trail of glossy black water in the green gloom under a hovering canopy of low tupelo, titi, and gnarled cypress trees.

They say it is about 240 miles from the Sill to the mouth of the river. People who fancy their prowess with a paddle sometimes travel the whole river by canoe. It takes them seven or eight days and is more a test of stamina and patience than a pleasant outing. The best canoe reaches are on the tributaries and from the Narrows down to Ellaville on the main river. Accessible launching and haulout sites are conveniently located. The upper forty-odd miles of the river run through pine flatwoods and hammock, and the clear black water slides over bars and slip-off slopes of white sand. Little water-table springs spread crystal-clear fans into the dark amber of the main stream. The water is very acid, and the fauna is lean.

Farther along, the river cuts into bedrock, and limestone bluffs and shoals appear. Down behind the big rapids, the jutting edge of the bluffs' more resistant strata have been honed to knife edge by the abrasive action of Lord knows what ancient raging waters. For long stretches the bluff face is lined, just above high-water mark, with a continuous rank of cinnamon ferns. In March and April wild azaleas crowd the shores.

The Suwannee has three main tributaries. The first is the Alapaha, which arises far up in Georgia and, in times of reduced flow, disappears into the earth and continues southward underground. It emerges in Holton Springs and Alapaha Rise, two of the larger springs of the Suwannee Valley. The last time I went out to the Alapaha it was only a fifty-mile stretch of dry white sand through the Georgia flatwoods and hammock land, and a man invited me to ride down the waterless riverbed in his four-wheel-drive vehicle. It seemed that half the county was out there doing the same thing. Downstream at Ellaville the Suwannee is joined by the Withlacoochee, which also comes down out of Georgia, staying above ground the whole way. The third main tributary is the Santa Fe, which originates in Santa Fe Lake in Alachua County, receives two tributaries, New River and Olustee Creek, and goes underground at O'Leno State Park. It reemerges after a two-mile absence. Twelve miles downstream from the rise it is joined by Ichetucknee Springs Run; and six or seven miles farther down it joins the Suwannee. Where the two rivers come together the Santa Fe looks like the main fork, and through the years many people, forgetting whether you bear right or left there, have gone the wrong way. The junction used to be a tranquil place, with deep holes where sturgeon jumped in summer and longnose gars slashed at the surface and a few people came to fish in little boats. It is now an arena for water-skiers.

In its upper section the Suwannee crosses the level land of the Northern Highlands physiographic region. Near Ellaville it cuts through an erosional scarp and moves out into the Gulf Coastal Lowlands. Every other stream that crosses the scarp between Gainesville and the Suwannee goes underground and then comes out again somewhere. Even tiny brooks do it. The Suwannee stays out of the earth, but on its way across the scarp it descends through a series of surprisingly strong rapids. Whitewater fanciers, who find little of their kind of country in Florida, make special trips out to the river above White Springs to run the "big shoals." Experts would find the whitewater there quite short and tame; but the last time I was out there I saw a yellow canoe standing up on its bow, and two girls and a dog momentarily airborne.

Below Ellaville, where the Withlacoochee comes in, the river is fed increasingly by big artesian springs. The black water of the upper reaches gradually fades, the floodplain spreads, and as the banks move apart, a canoeist loses touch with the sights and sounds of the shore. Fishing improves, however—that is, the diversity of expected fish grows. The upper Suwannee is the home of red-fin pickerel and redbreast bream, of quiet angling with cane poles. There is sophisticated grace in cane-pole fishing, and I always envy the people I pass, up there in the redbreast, red-fin pickerel reaches of the river. Their catch after a day's fishing may be meager by Florida standards, but it is far more decorative than a string from the lower river, and they have lengthened their lives in catching it.

The lower Suwannee is, in some years and seasons, sixty percent springwater. Because this proportion changes with the rainfall in the upper floodplain there is a three-way shift in the concentration of springwater, acid-black swamp water, and silt-bearing runoff from the eroding surfaces of the basin. When Bartram stayed at Talahasochte, down near Manatee

Springs, the river was in one of its clear modes. This is how, in *Travels,* he described the river one day in 1773:

> Having supplied ourselves with ammunition and provision, we set off in the cool of the morning, and descended pleasantly, riding on the crystal flood, which flows down with an easy, gentle, yet active current, rolling over its silvery bed. How abundantly are the waters replenished with inhabitants! The stream almost as transparent as the air we breathe; there is nothing done in secret except on its green flowery verges.
>
> Behold the watery nations, in numerous bands roving to and fro, amidst each other; here they seem all at peace, though, incredible to relate! But a few yards off, near the verge of the green mantled shore there is eternal war, or rather slaughter. Near the banks the waters become turbid, from substances gradually diverging from each side of the swift channel, and collections of opaque particles whirled to shore by the eddies, which afford a kind of nursery for young fry, and its slimy bed is a prolific nidus for generating and rearing of infinite tribes and swarms of amphibious insects, which are the food of young fish, who in their turn become a prey to the older. Yet when those different tribes of fish are in the transparent channel, their very nature seems absolutely changed; for here is neither desire to destroy nor persecute, but all seems peace and friendship. Do they agree on a truce, a suspension of hostilities? Or by some secret divine influence, is desire taken away? Or are they otherwise rendered incapable of pursuing each other to destruction?

I can find nobody who is old enough to have seen the Suwannee when it was as clear as Bartram described it. Long ago, however, I decided that looking for fabrication in Bartram's reportage is unrewarding. Once in a while he misinterpreted, but he almost never misobserved; and if he said the lower Suwannee was clear, it probably was clear. The big springs no doubt were flowing more freely in those days, and if Bartram's stay at Talahasochte coincided with drought in the upper basin, the preponderance of springwater could account for the lower turbidity. As for the fish in the clear channel living peaceably together, you see that all the time in any of the big springs in which fish have been allowed to remain.

Just below Fannin Springs the river becomes tidal, and the floodplain forest broadens. Originally there were stands of giant bald cypress here, but they were timbered off long ago, and in the second-growth swamps tupelo and other broadleaf hydrophytic trees are dominant. The cutting of the Suwannee cypress began more than a hundred years ago, and there are fabulous tales of the gigantic logs that were taken out. Their size is suggested by the oceangoing dugouts in which the Indians made their trading voyages across the Straits of Florida to Cuba. At the town of Suwannee, née Demory Hill, the river enters Suwannee Sound by East Pass and West Pass, each of which has numerous little distributaries. Along these creeks, tupelo and cypress are displaced by curious mangrovelike stands of corkwood—*Leitneria floridiana,* whose wood is lighter than cork and is used for floats by fishermen. The sound is margined by broad sweeps of sawgrass, black rush, and spartina that are set with little limestone islands wooded with cabbage palm, red cedar, live oak, and red bay. Nearly all of these palm islets are old kitchen middens, built up by Indians who seem to have eaten little besides shellfish—oysters, clams, scallops, whelks, and every other kind of snail in the sound, including fighting conchs. When the Faber pencil factory moved to Cedar Key more than a century ago it was because of the abundant red cedar on the islands and in the coastal hammocks. All the big old trees were cut down. Today, piecemeal cutting for fence posts continues, but the little cedars keep coming back, and the little islands they live on under the cabbage palms, set in marsh out to the sky, are a classic landscape.

The jewels of the Suwannee Valley are the rheocrene springs, the effusions of artesian water that come charging up out of the rock then linger in the crystal basins before moving away through the woods to join the river. These superb fountains are a unique natural asset of Florida. They are found nowhere else in such size and abundance. When I think back through the little gems of natural landscape that have stirred me, none had the dreamlike quality the springs had before they were found by the masses of people who admire and remodel them today. The springs I remember most vividly were the ones you traveled to down long, sand-track roads through dry pine hills. The first sign of something different was the dark green of a clump of broadleaf trees; then you saw the spring boil like a blue gem in its setting of green hammock, its water tumbling up out of its deep birthplace and roiling the surface with little prisms that sprayed color from the slanting light of the morning sun. Where the surface lay quiet, the deep places were some shade of blue, the current-washed bottom was snow white, and the rest of the basin was spread with polychrome gardens of half a dozen kinds of submerged water plants. The clarity of the water was absolute. You could watch a crayfish juggling a dead minnow forty feet down, or a stovewood catfish chewing water as he peered from his lair in the mouth of a side boil. But let William Bartram tell how Manatee Springs looked to him when he visited it while staying at Talahasochte:

About noon we approached the admirable Manatee Springs, three or four miles down the river.

It is amazing and almost incredible, what troops and bands of fish and other watery inhabitants are now in sight, all peaceable; and in what variety of gay colours and forms, continually ascending and descending, roving and figuring amongst one another, yet every tribe associating separately. We now ascended the crystal stream; the current swift: we entered the grand fountain. . . . The ebullition is astonishing, and continual, though its greatest force of fury intermits, regularly, for the space of thirty seconds of time: the waters appear of a lucid seagreen colour, in some measure owing to the reflection of the leaves above: the ebullition is perpendicular upwards, from a vast ragged orifice through a bed of rocks, a great depth below the common surface of the basin, throwing up small particles or pieces of white shells, which subside with the waters at the moment of intermission, gently settling down round about the orifice, forming a vast funnel. At those moments, when the waters rush upwards, the surface of the basin immediately over the orifice is greatly swollen or raised a considerable height; and then it is impossible to keep the boat or any other floating vessel over the fountain; but the ebullition quickly subsides; yet, before the surface becomes quite even, the fountain vomits up the waters again, and so on perpetually. The basin is generally circular, about fifty yards over; and the perpetual stream from it into the river is twelve or fifteen yards wide, and ten or twelve feet in depth; the basin stream continually peopled with prodigious numbers and variety of fish and other animals; as the alligator, and the manatee, or sea cow, in the winter season. Part of a skeleton of one, which the Indians had killed last winter, lay upon the banks of the spring. . . . The flesh of this creature is counted wholesome and pleasant food; the Indians call them by a name which signifies the big beaver. My companion who is a trader in Talahasochte last winter, saw three of them at one time in this spring: they feed chiefly on aquatic grass and weeds.

The hills and grove environing this admirable fountain . . . occasioned my stay here a great part of the day; and towards evening we returned to the town.

More poignantly nostalgic to me than even Bartram's impressions of Manatee Springs are my own memories of the magnificent spring-run landscape of Ichetucknee as recorded in my book *Ulendo: Travels of a Naturalist In and Out of Africa*. It was back in a time before tubing; before the Crackers started storming up the spring runs in boats with ten-horse Johnsons; before Cousteau perfected his first scuba regulator—a time so far back that the face mask through which I saw the things I told of was only a circle of window pane in a headpiece cut from an inner tube: "In those days it was enough just to ride down with the stream and look for things to see and never see a single beer can from Ichetucknee Springs to the Santa Fe. The sun was hot on my back through the wet lap of water and I hung belly down in the air-clear stream and looked at the bottom slipping by, one moment a waving yellow-green of *Sagittaria* ribbons and the next the black-green of a naiad bed or the sudden red of water purslane, or moving soft horns of pink-tipped coontail set with slim cones of spiny snails."[2]

It is a fine sight down through new spring water. One of the sad parts of my lot is that goggling needs more drama than pretty plants to stir my slight metabolic fires and keep off the dire sickness of the skinny skin diver—the ague we used to call the "big shakes."

There was that day, for instance, a long run with no break in the water gardens—no bone-strewn riffle or shards of Indian pottery or Suwannee chicken cooter shying at my passing—with nothing there but the man-killing cool and the sweetness of a spring run at summer noon. The zeal began chilling out of me, and I thought how welcome a fire on the bank would be, and a chocolate bar; and the bottom began to look like only wet plants. Then an eddy swung me over a bed of stonewort and I felt the prickle of the little leaves, and the smell made one of those queer smell-imprints that stick hard forever; so that now, to me, the faintest scent of stonewort or even of some sorts of onion soup brings back the sight of the big molar tooth lying there on swept sand beyond the stonewort bed, with all its roots and cusps, and its enamel still shining as if a big man had lost it the day before. It was the tooth of a mastodon, and the heat came back in me at the sight. I worked back along the edge of the sluicing channel, dived, and grabbed the tooth and rode the current down to slack water. Then I stood there knee-deep in more musk grass, turning the fossil in my hands, looking back plainly to a time when real giants lived in the Suwannee Valley land. The tooth was half the size of a football—too big for the bag that held my match bottle and chocolate bars—but its being so ponderous and such sure proof of different times had warmed me, and I dropped back into the current to ride on down, with the four-pound tooth held tightly in one hand. There was a quarter-mile of smooth travel, with now and then a Suwannee bass rolling his red eye from under a jutting log or a sprinkling of silver minnows all nose-upstream in the channel edge, or a moss-thatched stinkjim craning his neck to scramble from under my slipping shadow. Then suddenly I slid over another swept shallow and in a litter of meaningless pieces of brown bone saw bulk like a shaped lump of coal half out of the dark sand bottom. I clawed back to a point upstream from the object, dived, and rooted out the chunk of shiny black with my one free hand. Then I kicked away downriver to feel for shallow water and see what the new find was. When my feet found bottom, I gulped air, scratched the mask off, and stood studying the object.

It was clearly a piece of another tooth, but it was very different from the first. It had an undivided root, and the crown was crossed by repeated low, wavy ridges instead of high cusps. One end was broken off, and I was barely able to summon the lore to know that this, too, was a bit of a proboscidean, a relic of some elephant-kind of another sort. It was the grinding molar of the big Columbian mammoth. I thought the matter through, and then stood there in the suck of the current with bits of two Florida elephants in my two hands. I was so fired up by the coincidence and the triumph that the water stayed warm as new milk for the whole mile down to the landing.

Besides telling of a more tranquil time at Ichetucknee, that passage suggests the role of spring runs, and of the Suwannee system generally, in eroding out fossils and displaying them in their beds. In other regions stream riffles run over pebbles or cobbles. In the Suwannee drainage, where pebbles and cobbles are few, the fast shallow of a spring run is likely to be paved with bones, relics of the interglacial stages of the Pleistocene. Geologists of the Florida State Museum have spent hundreds of hours diving or dredging up fossils and archaeological relics. When scuba gear first came into use, and it was possible to go far back underground in the deep springs, divers found piles of alligator and turtle bones mixed with those of extinct elephants, camels, giant tortoises, and Pleistocene horses.

There are more artesian springs in Florida than in any other state—or in any other region of the world, for that matter. Of three hundred major Florida springs, about one hundred occur beside or within the channels of the Suwannee River or its main tributaries. They cluster there because the river has cut down through overlying sediments into the limestone of the Floridan Aquifer—the stratum that holds much of the water supply of the state. The limestone was laid down during Tertiary time, partly by chemical precipitation of calcium carbonate and partly by the accumulation of shells and other bits of marine creatures that swarmed in the warm seas of the time. The rock later was covered by sediments, then it warped up above sea level in a geologic event known as the Ocala Uplift. As the overlying sediments eroded, the limestone bedrock was subjected to the dissolving action of rainwater, which was made weakly acidic by carbonic acid from the air and organic acids from plant debris. As the water percolated through cracks in the limestone, the fissures and cavities enlarged, and with time the bedrock of the region became a honeycomb maze of interconnected tubes holding water under pressure. Wherever erosion cuts into these waterfilled conduits, artesian springs are formed. The water in them comes out under pressure because the conduits lead down from higher recharge sites. Water-table springs, on the other hand, merely move down through permeable strata at the level of the water table and seep out where these are cut. Florida has both kinds, but the artesian, rheocrene springs that boil up and flow away as little rivers are the singular blessing of the Florida landscape.

The big springs are not confined to the Suwannee Basin. The best known of them all is Silver Springs near Ocala, and its run is a tributary to the Oklawaha, which in turn flows into the St. Johns River. Wakulla Springs, in the Panhandle, south of Tallahassee, is the deepest of the Florida springs. It emerges at a depth of two hundred feet and joins the St. Marks River, which flows into the Gulf. Southward along the coast, other big springs—Weeki Wachee, Homosassa, Crystal, Chassahowitzka —form streams that flow a few miles into the Gulf of Mexico. Short distances off shore along both coasts other huge springs emerge from the sea bottom, some of them sending up drinkable fresh water with such force that it is hard to keep a boat in the middle of the boil.

I began making frequent visits to the Suwannee and the Santa Fe when I was an undergraduate at the University of Florida. I was taking Professor Leonard Giovannoli's course in herpetology and ichthyology, and the state list of reptiles and amphibians was just beginning to fill out. The Suwannee Valley is important as a transition zone separating endemic species and geographic races of the peninsula from the fauna of the continent to the north. So I spent a lot of time out there with Oather Van Hyning, son of the director of the Florida State Museum and a superb naturalist, trying to add to the state list of amphibians and reptiles or to extend the ranges of those already known. It is hard to explain how obsessive the ransacking of wild nature for new state records used to be for a youthful zoologist, back in those days before zoology depended on canonical analysis; before the coeds arrived at Gainesville; when, anyway, young women were a lot more parsimonious with their distracting favors; and when there wasn't even any beer. Collecting reptiles and amphibians and fish was a large and exciting part of life when I was an undergraduate.

One of my early herpetologic triumphs was coming upon a pair of brick-red water snakes copulating on a limb. I found them on the Santa Fe, just below the rise. When I looked down from the edge of a low bluff and saw the amorous snakes on a limb hanging over the water—and saw that they were quite clearly not anything I knew in the museum or on Leonard Giovannoli's spot quizzes—I crept along the edge of the bluff with all the stealth I could manage. Once alongside the snakes' limb, I jumped and grabbed with both hands, and when I hit the water I had the first two specimens of *Natrix erythrogaster* that Giovannoli and the museum curator had ever seen.

Another early reward the river brought me—a small thing, perhaps, but obsessed as I was with exploring for cold-blooded vertebrates, a joy—was finding *Eurycea bislineata* in a river birch swamp. At the time I was not aware that birch occurred in eastern Florida, so finding the swamp was stirring in itself. Then I went into it and started turning over logs—the dominant collecting technique of those times—and very quickly turned up a thin, astonished little nondescript salamander that was clearly not *Desmognathus* or *Plethodon*. I took it home and found it to be *Eurycea*, and for a short while I was a minor hero in my little circle. We had seen its kind out in the Apalachicola ravines but never as near home as this, and since that day I have never seen another in the Suwannee drainage.

One of the early biology courses that I took at the university was freshwater ecology. For that, Speed Rogers used to take us out to the Santa Fe at Poe Springs, where a boulder bar and a broken-up dam made a patch of fast water. Rapids are a rarity in peninsular Florida, and the rocks and riffles there were peopled with little creatures not common elsewhere. Under the stones there were larvae of six or seven kinds of current-loving mayflies, dragonflies, and damselflies; and there were belligerent hellgrammites and even stoneflies, which are almost unknown elsewhere in the river. Blackfly larvae clung by one end to the stones and vibrated in the little torrents, and caddis fly larvae built minute trap nets across the mouths of their anchored cones and fed themselves on whatever edible flotsam got caught in the webbing. There was a different planarian there, a little aquatic flatworm of a kind not found in the quiet acid waters I was used to. There were limpets, too, as I recall, and *Goniabasis* snails—long, thin cones set with spines to keep them from rolling away in the sweep of the stream. There were three kinds of darters in and around the rock pools in the rapids and two kinds of madtoms—two-inch tadpole catfish that stick your fingers with hot, painful pectoral spines. This array of creatures was among the finer things of life to Rogers. I liked them, too; but I remember the self-discipline it took for me to stay out with the class in the rapids, wielding a tea strainer and forceps, dumping tiny animals into an enamel pan and studying them with a hand lens, when it was the cold-blooded vertebrates of the river—the fish, amphibians, and reptiles—that stirred me most deeply.

As North American fish faunas go, that of the Suwannee and its tributaries is not numerically imposing. It includes some seventy species. Most eastern rivers have more, mainly because of their diversity in two groups of fish, the darters and the cyprinid minnows, in which the Suwannee system is lean. Two kinds of fish are endemic to the river. One is the Suwannee bass, a species more closely related to the Kentucky bass than to the largemouth, which also inhabits the river. The other is a banded pygmy sunfish that has not yet been described. Some of the Suwannee fish—the sturgeon and striped bass, for instance—are anadromous, but marine fish also wander into the Suwannee. Mullet, tarpon, gray snapper, needlefish, and bull shark all are able to tolerate changing salinity without physiological disaster, although why they wish to go up rivers is unknown. Of the confirmedly euryhaline fish, two seem straight out of Alice's Wonderland: the Gulf pipefish and the hogchoker. The pipefish is a straightened-out—or, better, unbent—sea horse; the hogchoker is a little sole. I first saw a hogchoker at Ichetucknee back before mobs of people had wrecked the basin of the main spring. Between breaths I was lying on the bottom and, through the pane of a mask, watching a big soft-shelled turtle that lay buried in the sugar-colored sand. It was only a low, sandy hump with a thin, tubular snorkel nose sticking out. Looking closely at the bottom I suddenly made out another, smaller hump on the rippled sand, and when I poked it lightly with a fingertip it erupted, dashed away for a yard, and embedded itself in the sand again. It was a two-inch sole, and my first contact with any freshwater flatfish. Actually the hogchoker is well known in other southeastern rivers, and it proved to be common in Ichetucknee Run. It was there, also, that I first saw pipefish in freshwater. They lived in beds of water plants, clinging to the thin leaves of *Philotria* and *Chara*. They never were abundant but usually could be taken in a little net if you swept it through the vegetation.

Of all the expatriate marine fish, the pipefish was to me the most bizarre and unlikely. I didn't know then that the sea horse group to which it belongs is well known for its invasions of streams. Not long after I learned about the Ichetucknee pipefish, Homer Smith, the eminent student of the physiology and evolution of the vertebrate kidney, visited the zoology department. Smith had just received a grant to go to Thailand to collect freshwater pipefish. I was almost able to refrain from telling him that pipefish also lived in fresh water in the Suwannee; but I was young and not able to resist the urge. Smith was taken aback, but only momentarily. "All right," he said. "That's great. We'll get to them when I return."

Another marine visitor that sometimes turns up in the river is the bull shark, the only kind of shark that habitually ascends American streams. Its visits to the Suwannee must be rare, but they are reported occasionally;

and there is an old account of one being caught at Branford, fifty miles upstream from the mouth. The bull shark is common in rivers along the Caribbean coast of Central America. There is even a flourishing population in Lake Nicaragua, whose only marine connection is the seventy-five-mile-long Rio San Juan. This colony has been recognized as a separate species, but a study by Thomas Thorson shows that bull sharks tagged at the river's mouth make direct journeys upstream, passing strong rapids in its middle section.[3] A Creole friend of mine once spent three months beside the San Juan rapids as watchman for a steamboat that was being repaired, and he often saw sharks in the rapids. They thrashed hard through fast patches, he said, then rested behind big rocks, then thrashed up to the next quiet place. So the sharks don't just wander aimlessly up the river, but what their clearly strong ecological motivation is, nobody knows. It is not a breeding drive. The only known breeding place is said to be in the open Gulf of Mexico, off the mouth of the Mississippi, and you never see any young sharks in the rivers.

My son Tom once caught a three-hundred-pound bull shark when he was working on our sturgeon tagging project. This was not in the Suwannee but in a deep hole in Brothers River, a tributary to the lower Apalachicola. Brothers is a little black-water creek with low shores of tupelo and cypress, with cane-pole fishermen usually sitting quietly in johnboats waiting for catfish and bream to bite. When Tom finally got the shark out of his sturgeon net, instead of stowing the shark properly in the bottom of the boat, he hauled it up onto the bow and arranged it so that the anterior third projected forward like a figurehead. I picture the effect as not unlike that of the P-40s of World War II, on whose noses the crews used to paint shark faces. Most of the people in the boats along the shore probably had been fishing there all their lives and maybe had never even seen a shark in the tranquil creek. Anyway, there was a mild sensation when Tom came by at dead-slow speed, giving everybody time to react in a proper way. One woman just lay down her pole, pulled in her little anchor, and paddled away. The reaction of a farmer in another boat was altogether admirable. He gaped briefly at the spectacle, then turned solicitously to his wife, who was still gaping, and drawled, "It ain't rale, Rachel. It's one of them plastic ones the kids blows up and floats on at the beach."

The sharks are quiet in the rivers. The sturgeon jump, and in June on the Suwannee, people even complain of being kept awake by the noise of their splashing at night. But the sharks are silent and self-centered, and there may be more of them there than we know about. In any case, Russell Bougard had a chilling experience with a Suwannee bull shark one afternoon. Russell lived at Shell Landing up on Holmes Creek, and my son

David and I went up there one day to talk to him about sturgeon. He fished for them commercially. His main calling, however, was salvaging sinkers—old sunken cypress logs—from the river bottom. He told us that he had walked farther along the bottoms of Florida's rivers than anybody alive. Most of the state's rivers are at least partly bordered by cypress, and back when the original timber was cut, logs sometimes sank to the bottom, staying there because cypress practically never rots. For fifty years and more, people had been searching out and raising these sinkers, and with cypress selling for $1,200 a thousand feet, any log that could be salvaged was worth working with. Most of the easy ones had been found; but in the deeper, darker reaches of the rivers some remained, and they were the ones Russell Bougard went after. He used to be a scuba diver and demolition expert in the Navy, and he spent several years working underwater in the Pacific. "I saw some pretty hairy sharks out there," he told us, "but where I got the living hell scared out of me was right here in Florida in the Suwannee River."

He was raising sinkers on the lower Suwannee and had found a big one just below Fowler Bluff. It was fifteen feet down in dark water, and he was trying to get chains in place to winch it up. He couldn't see far, he said, only a few feet, but as he worked he got a sort of feeling that something was taking shape in the murk above him. "It looked damn near as big as the log I was working with," he said. The shape kept looming, and slowly it became a shark; and Russell just about, as he said, peed his pants. He froze and stayed frozen for a couple of seconds, and the shark grew solid, rolled its eyes, then very slowly faded into the upstream gloom.

Another creature that gave the Suwannee a special glamour and brightened my student days was the Suwannee chicken. This animal is not poultry but a striped-headed cooter of a kind confined to the Suwannee River and adjacent Gulf coastal streams in Florida. Before I ever saw the Suwannee chicken I had heard the Cracker fishermen extol its eating qualities. "Best meat there is," was the unanimous verdict. "Way better than the regular streaked head you find in the ponds." And though nobody could say why, everybody agreed that the turtles from the Suwannee were better eating than their relatives in streams to the north and south.

That bit of ethnozoology gripped my interest, because I used to be addicted to eating wild animals of all kinds. I was even more excited when I caught some Suwannee chickens and found that they are an undescribed kind of turtle—a race most closely akin to river turtles of Georgia and South Carolina above the fall line but clearly different from anything else. Besides being different from all other cooters in markings and coloration, the Suwannee turtle has a smoothly streamlined shell and extra-broad back paddles, and it swims faster than any freshwater turtle I know. In personality, too, it is a thing apart. Though timid and skittish in the wild, in captivity it is less introverted than other cooters, craning its neck around to inspect new surroundings almost from the start. In maturity it is almost completely herbivorous, grazing on the copious submerged hydrophytes of the river and spring runs, and even moving out into the *Thalassia* beds in brackish water at the mouth of the river. At my stage of herpetologic development, finding this new turtle was stimulating, and presenting it to science as a new subspecies was a privilege.

At that time, even though the gastronomic cult of the Suwannee chicken was prevalent throughout the Suwannee Basin, people were fewer, and trapping pressure on the turtles was supportable. Nearly every old log or snag along the shore had its array of thirty or forty turtles or more. Even back then, however, they were very shy. It was hard to get within a hundred yards, and they often would start sliding into the water a quarter of a mile away. People caught them by tying fishhooks to short lines and fastening the lines to the top of a favored basking log; or by laying out a row of steel traps there; or by fastening net baskets along one side of the log and scaring sunning turtles into the baskets by approaching suddenly in a boat from the off side.

I ought to make it clear that though I used to eat Suwannee chickens with intemperate zest, I don't anymore. For a decade, at least, the Suwannee colony has seemed on the verge of disappearing. It is now protected by state law, but only halfheartedly—a family can take two turtles a day, and the regulation is obviously unenforceable. Unless it gets complete protection soon, the chicken seems sure to be lost from the Suwannee.[4] There are other kinds of turtles in the Suwannee, of course, and all are admirable; but for me, none has cast the spell of the Suwannee chicken.

Unless it was Jasper. Jasper was so sensational that telling about him sounds like a pack of lies. He was an alligator snapper from Jasper, Florida, on the upper Suwannee River. The alligator snapper is the biggest of all American freshwater turtles, and the Suwannee is the easternmost river in its southeastern range. Jasper weighed ninety pounds when we got him and more when he left us, though how much more is not known because he disliked being weighed. He was hooked on a cane-pole line in the Suwannee by the wife of a Jasper farmer. The lady immediately dropped the pole, but the turtle, instead of leaving, stayed there clawing at the thin line. After a while the farmer was able to drop a rope noose over the turtle's front leg and haul him ashore. The couple took him home in a pickup truck and put him in one of those big black iron sugar kettles that you see around Florida farmhouses—the ones that look like the steel helmets they used in World

War I. Quite a lot of people came to look at the turtle, and his picture got in the Jasper paper; then he was taken to the Halloween festival and exhibited as a prehistoric monster. By that time practically everybody had seen him, and the farmer telephoned the university to ask whether a turtle that big was any use to science. I heard about the offer, and though it was not exactly for science, I did want an alligator snapper to keep in our little fish pond; so my sons Chuck, Stephen, and Tom drove over to Jasper to get this one. When they got there they found the sugar kettle upset and the turtle gone; but he had headed out through a cornfield, and they had no trouble trailing him through the broken-down cornstalks and into a patch of briars. They brought him home.

We named him Jasper, and he lived for a year in the goldfish pond under the living room window. He was hand-fed two or three fish a week in warm weather, and he became beloved by the people in the house. One bleak day in spring he climbed out and disappeared into Wewa Pond, and he stayed out of sight for more than two weeks. But one day we saw his head sticking up twenty feet from the landing; and when we all rushed down waving mullets, Jasper came straight ashore and accepted the offerings just as he had before. For seven more summers, when anybody went down to the shore Jasper came up, Caliban-like, from the depths of the pond and accepted hand-held fish with astonishing decorum. A wide circle of admirers built up, and it was a sad time when he got a squirrel stuck in his throat and expired. To this day when a swirl of breeze tilts a bonnet leaf out on the pond and the silhouette looks like old Jasper's importunate head, a little surge of memory and a sense of loss come to people at our breakfast table.

Two traditional abuses of the Suwannee fauna have diminished. One is imbecilic shooting at turtles, alligators, and any other wild thing from boats traveling on the river. This still goes on, but it has slackened, partly because the animals are fewer and more unapproachable and partly because the boats travel too fast for shooting.

Another disruption of the Suwannee fauna that appears to be fading away is cut-bait fishing. In Suwannee River parlance, cut bait is dynamite, little chunks of it with fuses trimmed short to detonate just as the charge hits the surface where schools of mullet are swimming. Cut bait is very effective, but it is hard on the ecosystem, and it is not a very safe way to fish. The chief occupational hazard is loss of members. There used to be more missing fingers on the Suwannee, it seemed, than anywhere. When Frederic P. Wortman went down the Suwannee in a rowboat in 1912 he found cut bait in use all along the spring-fed reaches of the river and in the springs themselves. He said fishermen cut their fuses so short they would sometimes explode before the dynamite hit the water; and "when stirred by a drink or two, [fishermen would] stand and throw small charges at a tree so well timed that they would girdle it." At Barnes Ferry he saw a man who had cut his fuse even shorter. The dynamite had gone off in his hand, blown it off, and driven the finger bones into the side of his face, where they remained permanently. When I first began visiting the Suwannee and Santa Fe back in the 1930s, cut-bait fishing was still widespread. You could hear the thumping boom of underwater explosions all up and down the river. The practice was kept alive by the constant coming of new schools of mullet from the Gulf. Aggregations of the resident freshwater fish had long since been broken up, but the mullet kept coming, and the cut-bait fishermen kept after them. All along the river the fish for church sings and fish fries were provided by dynamiters, and a man gifted in the use of explosives was held in admiration. I say man—I never heard of a lady cut-bait fisherperson. Even then, the techniques had been illegal for fifty years or more, but enforcement was halfhearted. The most likely penalty was the loss of a finger or two.

On July 4, 1980, my wife, Marjorie, and I went to three places on the Suwannee that used to epitomize for me the grace of the river. We coldbloodedly went on the Fourth of July to learn how these once fair bits of earth were absorbing the shocks of overuse, which naturally reaches a peak on fine, hot Independence Days. The first place we visited was the fork where the Santa Fe meets the Suwannee. The once tranquil junction was a race course with speedboats and skiers tearing around in circles, driving the water out into the woods, blocking passage to a little flotilla of tired Boy Scouts in canoes trying only to continue their long journey from White Springs down to Branford.

The next place Marjorie and I visited that Fourth of July was O'Leno State Park, where the Santa Fe goes underground. There is broadleafed hammock around the sink and all the way down to the rise; and from there on, for a mile or more after the river comes out again, it flows through the forest in one of the loveliest half-mile sections of the whole Santa Fe. The forest was cut over a few decades ago, but Florida hammocks come back fast. The O'Leno woods looked good, and the park was standing up well under its Fourth of July burden. It was an anthill of people, but they were all in orderly arrays in stipulated places, not swarming indiscriminately through the woods or along the river but on trails and in the roped-off swimming area, the cookout zone, and the public toilets. A very small fragment of the protected area was providing outing space for thousands of people who were not wrecking the landscape that drew them there.

Our last and most apprehensive pilgrimage was to Ichetucknee Springs

State Park, and there it was different. I am one of many people who once considered Ichetucknee and its run the most beautiful landscape in the world. I had stayed petulantly away from the place for twenty years because too many people were going there. The first irritation had been a church group setting up some sort of field station in the woods beside the spring. After that the Boy Scouts started building merit-badge shelters and latrines all over the place. Then the swimmers began to multiply and by sheer hydraulic action devastated the main spring basin. Pretty soon snorkelers came, chased all the stinkjims and spring crayfish away, and even gathered up the snails. Then the scuba clubs discovered the Florida springs. The first ones came reverently to marvel at a new and unimagined world. The multitudes who come now see only vestiges of what their predecessors marveled at. Long ago they scared off the pallid cavernicoles and dispersed the myriad little catfish that swarmed in a big chamber back in Jug Spring. The spring entrances have been eroded by bodies and gear; the walls of the dim rooms have been white-grooved by scuba tanks. Some divers who come now would be just as happy in a system of culverts buried deep in the earth and filled with clear tap water. Others, I should add, deplore this lack of sensitivity.

And then finally the tubers arrived. Besides its other virtues, Ichetucknee Run is, of all the moving waters of Earth, the most idyllic tubing stream. On warm days tubers come by the hundreds; on warm holidays they are out there by the thousands. The stacks of tubes for rent along the highway from Fort White to the spring rise like the foothills of the Andes. Jeanne Mortimer, then a graduate student of mine, collaborated in a study of tubing use that the Department of Natural Resources was making in 1979. She sat all day in a bush by the run and counted the people who passed on tubes. Steadily the tubers slid by, singly, in loving couples, or in little clots, committing no overt violence on anything—just never ceasing to go by. The survey found the average prime-time population to be three thousand tubers a day. The magic of the loveliest waterway in the state is being eroded by the joyous, innocent, incessant passing of people on old inner tubes.[5]

Up to now the Suwannee has survived its abuses fairly well and has emerged as one of the least spoiled rivers in the eastern United States. The physical substrate has not been wrecked—or rectified; the water has not been badly poisoned or taken away. The biological landscape is to a degree self-healing. The Indian farms and Spanish mission sites have long been lost in woodlands that look far better now than they looked a few decades back, when scars were new and hogs and ground fires roamed everywhere. The grand old cypress trees were cut down, but their descendants have come back, and along with the ogeechee limes they form pleasant, swamp-shore stands. The springs have been brutalized; but even they could be restored with careful tending. The fauna is less abundant today, but its diversity is not much changed. Bachman's warbler and the ivory-billed woodpecker are no longer there; but I can think of no species lost from the Suwannee that has not been lost everywhere. Otters and alligators are growing hard to find, and the old alligator turtle is in trouble. Fish and migratory ducks have diminished sadly in the estuary. But the sturgeon jump on schedule, the silver-gold tarpon roll in the passes in June, roseate spoonbills wander in and decorate the delta, and more sea cows come into the river than have come for a hundred years. If serious saving were undertaken, there would be plenty left to work with.

Suwannee River Sturgeon

I first heard of sturgeon in the Suwannee River at Odlund Island, at the mouth of the river off the town of Demory Hill, recently renamed Suwannee. In those days Demory Hill was about as isolated as any town in Florida, except maybe Flamingo, Chokoloskee, and Steinhatchee. To get there we traveled twenty-five miles from the highway at Old Town through wet flatwoods on a road that was two wheel tracks except where it completely disappeared under the coffee-colored water of the bayhead creeks. I remember how pleased we were to have made the trip in only four hours. The town had the usual desultory clutter of a mullet-fishing settlement of those times. We went down there looking for information on the Suwannee chicken turtle, which I had just decided was a creature unknown to science. I wanted to learn about its life history, and people up the river told me the Odlunds knew more about it than anybody else. Besides that, Mrs. Odlund was celebrated for her cooking, and her most widely acclaimed dish was Suwannee chicken stew. Almost from birth I have been peculiarly tormented by Jekyll-and-Hyde compulsions both to learn about the natural history of animals and to eat them, so to me, Odlund Island had become a sort of mecca.

We found a boat to take us out through the river-mouth marsh creeks to Odlund Island. It was a pleasant place in the Gulf-island mode, a patch of high ground in the marsh, wooded with live oak, red bay, palms, and red cedar. Even back then the Odlund house had the comfortable air of having long been lived in by well-adjusted people of sense and tranquility. It was shady there, I remember, and a Gulf breeze rattled the cabbage leaves. There were red geraniums in cans on the veranda and dogs and cats walking amicably around. There were some shrieking jaybirds in the live oaks, and the redwing blackbirds were singing in the marsh. A male anole kept showing his throat fan on the banister, and a hyla surprised me by

Adapted from *Audubon*, 85, no. 2 (1983). Copyright © 1983 by the National Audubon Society, 700 Broadway, New York, N.Y. 10003.

calling from down in a cistern—or not by calling, by being there. When the Odlund children learned what I wanted, they started talking, not just about Suwannee chickens but about the half-grown green turtles and ridleys they caught on the turtle-grass flats in the sound, and about green-tailed snakes in the black mangroves—and then about sturgeon.

The talk of sturgeon was a surprise. I knew Oscar Odlund was involved with green turtles and Suwannee chickens, but nobody had told me he was a sturgeon fisherman.

Odlund was Swedish. He was a seaman in the merchant marine, but the flourishing sturgeon industry at Apalachicola had lured him ashore. He introduced the technique of drift-netting for sturgeon, which had never been used at Apalachicola. He soon heard that sturgeon were plentiful down in the Suwannee River, and in 1912 he moved there, married, and settled on his beautiful little island.

The many virtues of Odlund Island—the good fishing for largemouth black bass, channel bass, and speckled trout, combined with Mrs. Odlund's cooking, especially her irresistible Suwannee chicken stew—caused the family to be besieged by people craving lodging. With time a small, select clientele grew up. It was composed mainly of doctors from Atlanta, as I recall, and many of them grew old visiting the island every year, fishing and eating Suwannee chicken stew. When Mrs. Odlund died, Margaret Odlund Ghiselin, who had inherited her mother's cooking skill and had embellished it with sophisticated outside lore, took over the operation of the island retreat. The last time I saw Margaret she baked us a big black bass, stuffed with blue crab meat, and told us that a vivid memory of her childhood was of her father rowing back to the dock with the biggest sturgeon he ever caught tied beside the boat and, to her eye, about the same length.

From the time of that first visit to Demory Hill until 1975 I thought little about sturgeon in Florida. But one day John H. (Ben) Phipps, then president of the Caribbean Conservation Corporation, asked if I would be interested in organizing an investigation of the ecology and status of sturgeon in the Apalachicola River. That was a time when a few people were beginning to realize that the Gulf sturgeon, *Acipenser oxyrinchus desotoi*, was growing scarce and was probably worthy of endangered species status. Moreover, special interest groups in Georgia and Alabama were goading the U.S. Army Corps of Engineers to improve the Apalachicola channel so more soybeans and peanuts could be barged down out of the its two main tributaries, the Flint and Chattahoochee Rivers. Ben Phipps said that the Phipps Florida Foundation might furnish modest support for a sturgeon project. At the time, the closest I ever had come to seeing a live sturgeon was up in the Savannah River swamp, when I was very young. To get shut of me while he hunted turkeys, my father put me in a bateau with a half-black, half-Choctaw Indian. We sat in an eddy below a willow island for five hours while the Indian held a window-cord line with a weighted treble hook and waited for the line to be "rubbed" by passing sturgeon. No sturgeon rubbed, but from that day on sturgeon lurked in the back of my mind as elusive animals that I would like to know better someday. So in spite of my ignorance I gladly accepted Ben Phipps' offer.

The first thing we did on the Apalachicola was to travel the length of the river and talk with people about sturgeon. At every landing and riverside town we inquired about the sturgeon's abundance, seasonality, and habits, and about possible sources of fish for tagging and release. At Apalachicola, Joe Taranto, the one remaining shipper of roe for caviar, assured us that we could buy all the male and immature fish his fishermen caught. We learned that sport fishing for sturgeon went on just below the Jim Woodruff Dam at Chattahoochee and in some holes in the river above Blountstown. This was not bait fishing but the kind of "snitching" the Choctaw I spoke of did, with unbaited treble hooks. At Chattahoochee you anchor in a deep, hard-bottomed hole or downstream from a bar, lower a big weighted three-pronged hook, hold the line taut, and wait. When you feel a fish rub the line—which for some reason they love to do—you jerk, and with luck snag him in the belly. This exercise was much in vogue in the early 1970s, and because for some reason the Apalachicola snitchers rarely ate the sturgeon they caught, they seemed a promising source of supply. Besides tagging, we planned to search the river and its tributaries for evidence of spawning—courtship and mating or the presence of eggs, post-larvae, and fingerling sturgeon. The U.S. Fish and Wildlife Service was just beginning a survey of the anadromous fish of the river—especially the striped bass, but also the Alabama shad and the sturgeon. It was building pound nets at the mouth of the river and was planning to explore the upper reaches with seines and electrofishing gear and to set out plankton nets for larvae. The service invited us to join the project and take responsibility for the sturgeon.

We embarked with terrific optimism, but there were setbacks from the start. Joe Taranto, our friend and principal hope in the tagging work, died; and exceptionally high water kept the other Apalachicola sturgeon fishermen off the river. The flood hindered our netting in the lower river and creeks and the Fish and Wildlife Service's spawning-ground search in the upper river. Another bizarre reverse was caused by marauding alligator snapping turtles—which reach weights of one hundred pounds or more in the Suwannee and Apalachicola—that entered the government pound nets to get fish, then tore big holes in the mesh when they wanted to leave.

The snitchers failed us, too. The Blountstown people quit fishing because the town council made them stop throwing the dead sturgeon on the city dump, thus causing a public nuisance. This meant the snitchers had to bury the fish somewhere, and that was a lot of trouble and marred the sport. Some of the snitchers up at Chattahoochee told us that after the flood the fish just didn't rub anymore.

    The outcome of all this has been that, to the present, none of these activities has been renewed. The other day I talked with Anthony Taranto, Joe's son, and he said his fishermen had caught only one sturgeon during this spring's run. It was a two-hundred-pound female with twenty-five pounds of roe. Anthony said that nobody in Apalachicola nowadays knows how to prepare roe properly, and he wondered if I did. That suggested to me that Joe Taranto's skill at curing the eggs in a way that was acceptable to the New York caviar people might explain why commercial fishing for sturgeon at Apalachicola lingered as long as it did. When we left the river we were coming to realize that its sturgeon population was drastically depleted.

    The loss of sturgeon in the Apalachicola apparently progressed in two stages. One was evidently synchronous with a nationwide decline in sturgeon during the late 1800s. It was then that immigrants from northern and central Europe spread through the country and brought with them a strong predilection for sturgeon meat and roe. Within a few decades American sturgeon diminished everywhere. In Tampa, for example, where a thriving industry caught two thousand fish in 1886, only seven were taken in 1896. Today no industry is there at all, and it has been decades since anybody saw a sturgeon in Tampa Bay. When sturgeon grew scarce, avenues to markets from the more remote fisheries closed, and it was only the premium price paid for the preserved, easily shipped roe that sustained commercial sturgeon fishing. When Joe Taranto reached the river fifty years ago the fishery was already dwindling. He kept the industry alive because he knew how to cure the roe and find markets for it in the North. When our plans for the Apalachicola began to fall apart we started searching for a substitute site where by netting, tagging, and interviewing fishermen we might learn some helpful things about the biology of *desotoi*. Oscar Odlund told me he thought this could be done in the Suwannee. At almost the same time I got a copy of Alan Huff's paper on age and growth in Suwannee sturgeon, a publication of the Department of Natural Resources.[1] This clearly indicated that enough commercial fishing was going on to provide fish for a tagging program. We wrote to Ben Phipps of our troubles on the Apalachicola and asked his permission to move to the Suwannee for a time to learn to catch and tag sturgeon and to amass data that probably would be directly applicable to the Apalachicola situation if things ever picked up there.

Ben agreed, and, after arranging to buy Suwannee sturgeon from the nets and tether stakes of a local fisherman, we started tagging. The next season we bought our own set nets and put them out in the passes; then we learned the more advanced art of drift-fishing, which Odlund had brought down to the Suwannee from Apalachicola. In set-netting you find an eddy along the shore, stretch your big-mesh net from the shallows out through the eddy, anchor the outer end, and wait. The place to set nets on the Suwannee is in the passes, where the fish come in from the Gulf. Drifting is done farther upstream. The aim is to intercept fish moving up or down the river by keeping a one-hundred-yard net stretched out perpendicular to shore. To work the net you tie the outer midstream end to your outboard-powered boat and exert just enough pull to keep the net stretched across the current. You do this at night; and it can be either very boring or very exciting—depending on whether nothing happens, there is a bad lightning storm, you have to release a big tarpon from the net, or you have to tag and measure a one-hundred-pound sturgeon in the rain.

Oscar Odlund told me that if you set a small-mesh trammel net on the flats at the mouth of the river, especially in the fall, you could catch very small sturgeon—those weighing less than three pounds. We tried that, too, and soon found it to be true. After losing a lot of data in our early weeks on the Suwannee because the commercial spaghetti tags we were using fell apart, we began wiring Monel metal turtle tags to the thick, bony scales of the fish's dorsal keel. Those stayed on well, and during three seasons we were able to put them on 253 Suwannee River sturgeon. We had forty-four recoveries during one season in which the tags were put on; and there have been fifty-nine very interesting recoveries in later seasons. One sturgeon tagged in East Pass on March 23, 1977, at a weight of twenty pounds and total length of forty-four inches, was caught again on March 20, 1981. It was fifty-six inches long when retaken, and it weighed forty-five pounds. Besides gaining twenty-five pounds in the four-year absence, it returned to within two hundred yards of where it was first caught.

Nobody appears to have tagged southeastern sturgeon before, and we have been able to learn some new things about *desotoi*. A useful overlap of results from the Suwannee and Apalachicola studies is evidence of the fish's home-stream tenacity. Two kinds of evidence attest to this: only one sturgeon tagged in the Suwannee has been recovered in any other river, and our recovery percentage in the home stream is exceptionally high, as such tagging programs go, except for one instance. The only river in which sturgeon tagged in the Suwannee have been recovered is the Suwannee itself. This, of course, doesn't mean the fish never leave the river—they almost certainly do, though the evidence is wholly circumstantial. What it means is that the Suwannee is their only breeding place. If they conform to the pattern of the closely related Old World species, they range widely in the sea between breeding seasons. But, like salmon, they usually return to a specific home river—the one in which they hatched—to spawn. The Russians are beginning to learn where their sturgeon go in the Baltic after they leave the rivers, but in the United States we know little about this. Our tagging data provide no clues in this puzzle. The home-river fidelity our fish seem to show suggests, though, that protection and management of the population of *desotoi* in a given river would not be vitiated by its wandering away to other, less favorable rivers to breed. It also suggests that, as with salmon, the homing cues are imprinted in young fish; if so, transplanting hatchery-produced young might rehabilitate the colony of a depleted river.

On the average, fish recovered after a year have gained thirty percent in body weight. Little or no growth was shown when they were recaught during the same season. Some even lost a little weight. So *desotoi* does its growing in the sea.

This does not mean that no feeding is done in the river. I saw a man fishing for sturgeon at Branford—with a rod and reel, using "a wad of worms" for bait—and catching them, he said, once in a while. Our only other information on feeding has come from the stomachs of fish caught near the mouth of the river. Most of the sturgeon butchered by the fishermen are tethered—tied by their tails to stakes along the edge of the river—for up to three days before they are killed. A surprising feature of the stomach contents was the frequency of the brachiopod *Glottidia pyramidata*. Brachiopods belong to a class of primitive shelled invertebrates even older than sturgeon. They are excessively rare in most of eastern North America. The regularity of their occurrence in the diet of early-season Suwannee River sturgeon suggests that they are a staple food for the migrants as they approach the river mouth for the upstream journey. Evidence from stomachs, and from word-of-mouth reports of fishermen, shows that they are eaten by all sizes of sturgeon, but only as they arrive during the spring migration. The only other frequently found food remains have been fragments of little crabs and some green algae. On the Apalachicola, fisherman Sylvester Tarantino catches sturgeon on sea lice.

The tags that we recovered the same season in which they were put on told us some things about the movements of the fish in the river. Fishermen agree that the sturgeon begin to arrive in March, that they travel upstream to spawn, and that they start leaving the river for the sea in October and

November. All this was supported by our tag recoveries. But there is more to the pattern. For one thing, all the fish don't come into the estuary and charge straight up the river. Sometimes they move up in stages, stopping along the way for a few days or longer. We found that when sturgeon were retaken in the lower river, those weighing less than twenty pounds were more likely than the larger ones to be taken near where they had been tagged. This could mean that at their age their anadromous drive is incompletely developed, and this makes one wonder why they come to the river even though they are not going to breed. It is hard to believe that the young ones make the breeding migration as members of big schools of all sizes of fish, all heading for the Suwannee, but that only the big ones have any biological purpose in going there. The habit of remaining in the lower river is strongest in the very young fish, those weighing from one to four or five pounds. These are sometimes taken in mullet nets far out on flats adjacent to the passes, and we found little evidence of their presence anywhere upstream. Some people believe they are young of the previous year that have never left the Suwannee, but their strongly seasonal occurrence in that much-netted area makes a year-round stay seem unlikely.

This tendency of the younger age groups to linger in certain parts of the river was confirmed graphically by results from our tagging in the Apalachicola. We had not done at all well there, but in 1976 my son Chuck finally caught ten sturgeon up in Brothers River, one of the swamp-shored tributaries to the lower Apalachicola. On one set in a big hole there he caught four subadult sturgeon, tagged them, and released them a half-mile downstream from the capture site. Two of these fish were retaken the next night, having traveled the half-mile back up the river in about twenty hours. This suggested active site-tenacity, and subsequent results confirmed this dramatically. The two recaught sturgeon were again released, this time two and a half miles down from the capture site, outside the mouth of the creek and out in the current of the Apalachicola. The following night one of these was retaken back up Brothers in the same home hole. It had negotiated the fast-flowing main river, found the Brothers' mouth, and made a two-and-a-half-mile trip home, passing en route the mouths of two lesser streams, Little Brothers and Harrison Creek.

The place where these sturgeon were taken was a hard-bottomed hole. Angelo Fichero, a commercial fisherman in Apalachicola, told Chuck he had taken sixteen sturgeon there one night in 1972. He said he never found food in the stomachs of the sturgeon taken there, so feeding resources must not have been the attraction. What the appeal was is unexplained, but the tendency for immature fish to tarry in a restricted part of the stream agrees with our observations in the Suwannee.

The most discouraging setback in our sturgeon project has been the failure to find their spawning ground. We know they are anadromous. We know that big, sexually mature males and females arrive in the river in spring with well-developed milt and roe, and that when they leave in the fall they are spent. We know places in both the Apalachicola and the Suwannee where big sturgeon gather and jump into the air. But with the exception of a single post-larva taken by the Fish and Wildlife Service at Chattahoochee, no eggs or larvae have ever been seen by us or anybody we have talked to.

What is even more puzzling is that finger-size sturgeon likewise have never been found in either river. When they reach that size you expect at least to find people who are familiar with them. We have talked about this problem in hundreds of interviews from the town of Apalachicola up to the Jim Woodruff Dam, and from the town of Suwannee up to Fargo. Many of the people we talked to were familiar with sturgeon, but none knew where they breed, and none would even encourage us with lies about young ones of finger size.

One day up at Dowling Park, on the Suwannee, we seemed to be on a hot trail. My son Stephen and I had stopped at a little grocery store by the river and were doing our customary canvassing, going around from one customer to another and inquiring about the size of the smallest sturgeon he or she had seen. One man held his hands up around two feet apart, but the rest either had never seen a little one or had never seen any sturgeon at all. It was just the same as everywhere else; and it was discouraging, because at Dowling Park you are getting into narrow, faster, rockier parts of the river, with the kinds of habitat in which sturgeon elsewhere are said to spawn.

Finally we had talked the matter out, and everybody was standing around looking politely crestfallen about our problem, so Stephen and I thanked the people and started to leave. As we crossed the veranda an old man in a wheelchair at the far end of the porch yelled, "Hey, them fellers in there says they's not no little sturgeon in the river? Where the hell they think the big ones come from?" He cackled derisively, then shifted the wad of tobacco in his cheek, spat out among the sand spurs, and started talking about how he was the only man around who knew anything about the river and how he had fished on it for fifty-five years and could tell us anything we needed to know about sturgeon or any other kind of fish. We reminded him it was baby sturgeon we were looking for, and he said hell, yes, there was plenty of them, all the way from Dowling Park to Ellaville, and if he could just get down to the river he would show us some. We asked how big the ones he could show us were, and he said all sizes, down to the size of a starheaded minnow, which is maybe two inches long. He said you could catch them in a dip net. In spite of ourselves, we began to hope. But then we asked him what time of year they were there, and he said all the time; about then the storekeeper showed up at the door and beckoned, and when I went over he made derogatory motions toward his head and said, "That old feller's from the old folks' home across the road. He's in his second childhood."

So we walked away despondently. As we got to the truck the old man yelled that if he could get his damn chair down to the river he would show us some little sturgeon. We thanked him and drove off in low spirits.

It seems likely that we will not find the spawning ground until we use radio or sonic tracking gear to follow mature fish, taken at the mouth of the river in the spring, all the way to their upstream breeding places. We had a good gear-testing spree with a single big female sturgeon. On March 26, 1978, we caught a ninety-pound female at Long Reach on the lower Suwannee. We measured, weighed, and tagged her and wired a sonic transmitter to the base of her dorsal fin. The aim of the exercise was just to try out some borrowed tracking equipment. We released the fish where she had been caught and for ten hours kept in touch with her with a directional hydrophone. During the whole time she moved steadily upstream. At 2:00 A.M. the next day it started raining hard, and the trackers tried to net the fish and recover the transmitter. In the process, three other sturgeon that evidently had been swimming in company with her were caught, but she eluded the net. The party returned to Gainesville, frustrated.

Four days later a group tried to relocate the fish. On the somewhat naïve assumption that she would have continued upstream at the same steady rate at which she had moved during the first ten hours, we multiplied that by the elapsed time and decided that the place to search was between Hart Springs and Sun Springs. The trackers went there and, after prospecting for half a mile, located the fish at 12:15 P.M., just south of Sun Springs. Shortly after her signal was picked up, she stopped her upstream travel, began moving irregularly, then headed back downstream, stopping repeatedly to zigzag across the river. Finally, at 3:15 P.M., the fish entered a deep rocky hole, where she weaved and circled for more than two hours. At the same place, two other sturgeon repeatedly jumped into the air. At 5:15 P.M., our fish moved upstream again, and at 6:40 P.M. the tracking party had to leave the river. Two days later a six-hour search of the most likely sections of the river failed to pick up the signal. The overall distance traveled by this fish was about thirty miles, nearly all upstream.

The upshot of this incomplete but exciting little experiment was to add weight to what we already knew—that sturgeon are anadromous. The fish did not, however, reveal where the spawning ground is. With all our later searching, that remains a total mystery.

It is not completely clear why the sturgeon population decreased so drastically in the Apalachicola during the past two or three decades. The change in population level there is real, however, and it appears to be progressive. The most logical explanation seems to be that man-induced ecological disruptions have blocked the reproduction of the fish and obliterated essential habitat while exploitation has continued. The Jim Woodruff Dam at Chattahoochee cut off access to the Flint and Chattahoochee rivers thirty years ago. Although spawning in those streams was never documented, it almost certainly occurred. Sturgeon still congregate in diminishing numbers in some rocky holes just below the dam, and the single bit of direct evidence of spawning in either the Suwannee or the Apalachicola was that eight-millimeter post-larva taken in the Fish and Wildlife Service plankton trap set below the dam. So the dam probably has been a serious obstacle to reproduction. And now other vital sturgeon habitat in the Apalachicola is being demolished by the effort to maintain a nine-foot channel for barge traffic. In this exercise every habitat that sturgeon, at one stage or another of their sojourn in the river, seem partial to is being modified or obliterated. Rocks, snags, boulder bars and riffles, and rock shelves are being clawed away. Dredging and spoil disposal have changed the configuration of the river. Human remodeling of the Apalachicola is one easily identified factor in the recent decline. When this ecological disruption is combined with commercial and sport fishing—in both of which the prizes are the big, sexually mature sturgeon—it may not be necessary to look any further for causes of the loss. There also appears to have been a recent rise in the incidental take of sturgeon by boats trawling in Gulf coastal waters in February and March, but its importance is unknown.

Whatever factors threaten the future of the Apalachicola sturgeon, the population can probably be saved. If habitat destruction were stopped and a moratorium on commercial fishing and recreational snitching declared, recovery could almost surely be accelerated by releasing fingerlings from hatcheries. In Russia and California sturgeon hatcheries are operating successfully. None has progressed beyond the experimental stage in the Southeast. Within the range of the Gulf sturgeon, the Suwannee River population appears to be the most stable, and one hope for our sturgeon project was that it would reveal enough about the life cycle of the fish to justify setting up a hatchery there. Meanwhile, the taking of sturgeon weighing more than twenty pounds ought to be stopped and the search for the spawning place intensified. The failure of our search to date is dismaying, but it ought to continue. It should rely on tracking sexually mature fish caught in the lower river in April and on setting out plankton traps in likely places, including holes where sturgeon are seen jumping consistently. Meanwhile, Russian advances in inducing ovulation and milt production by hormone injections ought to be studied.[2]

Whatever is done should be done very soon. The prices the fisherman gets for sturgeon have more than doubled in the past two years, and there is no longer an uncertain wait for a truck to come by and take the catch away. Local fish houses now receive and freeze the fish, which trucks take to smoking plants in Miami. Interest in learning to prepare roe for the caviar trade is growing, and once the local people learn the technique, fishing will increase further. The pressure on the Suwannee sturgeon is sure to grow apace, and the Apalachicola population could disappear.

I want to say that I have formed a strong personal attachment to sturgeon. I admire their massive ganoid armor, and the ancient look of their heterocercal tails, protrusible mouths, and long dangling barbels. They are peace-loving, agreeable animals. No fish is so calm when brought into a boat. If sturgeon had the temperament of some other fish, our tagging project would have yielded less information and more injuries to personnel. Even a big bull sturgeon, with a hatchet-edged caudal peduncle more lethal-looking than a crocodile's tail, rarely thrashes about when brought into a boat but waits quietly and rolls his eyes. Sturgeon even remain calm when dogs lick them. My research assistant, Newt Meylan, has the bad habit of taking a big black German shepherd out in the sturgeon boat. Her husband, Peter, is never very happy about this, but he is indulgent, and the dog goes along a lot. When they bring a sturgeon up out of the net to measure it, the dog licks it from head to toe, as it were. This can't be reassuring to the sturgeon, but they never seem irritated; they only roll their eyes some more. I think the vast long time sturgeon have been on Earth has bred into them this admirable tendency of taking things as they come.

The fossil-like look of sturgeon is reinforced by their tendency to arrive at the mouth of the Suwannee in the spring having fed on brachiopods, members of an almost extinct group of shelled invertebrates that were abundant in the sea three hundred million years ago. It is stirring to think of these two living-fossil creatures coming down together from the dim Devonian, one eating the other, since a time when brachiopods were dominant in the seas and sturgeon were the most modern fish, and in all the world there were no land animals at all.

If the Tellico Dam was held up for the sake of *Percina tanasi,* what might not be accomplished with the sturgeon? The sturgeon is a super snail darter. He is big, venerable, amiable, even edible, and he cannot abide having rivers changed around.

An *I*ntroduction to the *H*erpetology of *F*lorida

Since the days of the earliest explorations, the herpetological fauna of Florida has evoked spirited comment. Hardly a mosquito-bitten Spaniard writing home for supplies or a French sea captain recording in his log the adventures of a shore-party but mentions "vipers" or "crocodiles," or the shocking noise the frogs made, or the Indian who tried to feed him snake. Colonizing Florida was such a strenuous matter, however, that zoological observation was considerably tainted with emotion, and only those forms of life which bit people, or which people could eat, elicited any enthusiasm in the early reports. And since the colonizing is mercifully not quite complete, current reports, too, retain enough of the old emotional taint to warm the soul of any decent herpetologist. Folks still tremble in the cabins when the satyriac bull-gator bellows on Middle Prong; and I know a man in the Scrub who will show you, without a trace of guile, the skin of a twenty-foot rattlesnake that he got from the boy who killed it and dragged its rattles off behind his horse.

Probably the first published illustration of a North American reptile is the alligator drawn from life by Jacob Le Moyne in his "Indorum Floridam provinciam inhabitantium eicones" (1591).[1] Le Moyne discloses that the Florida "crocodiles" grow longer than those in the Nile. The specimen figured is at least thirty feet long if we are to regard the gang of savages engaged in thrusting a pine trunk down the creature's throat as other than a race of pygmies. Le Moyne also depicts a culinary scene that includes the grilling, whole and unskinned, of a large snake of indeterminate species.

The earliest reliable observations on the reptiles and amphibians of Florida are those of William Bartram, which were published in 1791.[2] Although primarily interested in plants, Bartram was fascinated by the diversity of the southeastern fauna, and he commented with pertinence on all that he saw. His discussion of the Florida frogs, lizards, snakes, and turtles

Adapted from *A Contribution to the Herpetology of Florida,* Biological Science Series, no. 3 (Gainesville: University of Florida, 1940).

includes much accurate information on habits and habitats, and in most cases leaves little doubt as to the identity of the forms described.

From the beginning of the Seminole War in 1836 until the appearance of Edwin D. Cope's "On the snakes of Florida" in 1888, a number of popular but more or less inaccurate accounts of the snakes were presented by the authors of stories of Everglades explorations and of hunting and fishing in South Florida.[3]

Cope's essay later was incorporated into his *Crocodilians, Lizards, and Snakes of North America*, published by the Smithsonian Institution in 1900. The most important work of this period was that of Einar Loennberg, who spent several months collecting in the peninsula and published an annotated list of the sixty-six species that he observed.[4]

In 1910 Clement S. Brimley's "Records of some reptiles and amphibians from the southeastern United States" was published.[5] This paper included records of sixty-four species from Florida, several of them not mentioned by Cope or Loennberg.

Many of the most detailed accounts of the reptiles and amphibians of Florida are scattered among monographic revisions that treat various groups represented in the Florida fauna, short papers in which many of our species have been described, and published notes giving the first Florida records of various species.

The reptile and amphibian fauna of Florida comprises 162 species and subspecies belonging to 74 genera. The most conspicuous element is that which has invaded the state from the adjacent southeastern coastal plain and which consists of northern and western forms and coastal plain endemics. In addition, there are twenty-four Floridian races of more northern species, eleven endemics, seven West Indian species, and six more or less cosmopolitan forms (the marine turtles and geckos).

The most extensive invasion of Florida by the northern element is in the portion of the Panhandle that is drained by the Apalachicola River. The marked concentration in this area may be attributed to two chief causes. In the first place, the Apalachicola is the only Florida river that penetrates the continent as far as the fall line.[6] Its two tributaries—the Flint, arising in the Piedmont of northern Georgia, and the Chattahoochee, whose headwaters drain part of the red hill section of Alabama—have served as highways for the traversal of the broad coastal pine barrens by a number of species which under other conditions could probably never have extended their ranges south to Florida. Moreover, a short distance below the junction where the two rivers form the Apalachicola, the red clay hills of Jackson and Gadsden counties and the deeply dissected terrain of western Liberty County afford a peculiar and disjunct series of environments that are strikingly favorable for the accommodation of northern immigrants. The ravines cut into the bluffs along the east shore of the Apalachicola River, and in the clay hill section they are often deeply shaded, humid, and perpetually cool, with numerous small springs that give rise to clear, cold brooks that flow over rocks or gravel.

Many northern plants and animals reach their southernmost limits of distribution in these ravines. Beech, trillium, blood-root, and other species common in northern forests form a considerable element in the vegetation and confirm the northern aspect of the physiography. Numerous insects characteristically found in their aquatic immature stages in mountain streams occur in the ravine bottoms—many of them in flourishing abundance.

The Apalachicola ravines are noted for the amount of endemism that they exhibit. In addition to several plants—notably *Tumion taxifolium* (stinking cedar) and *Taxus floridana* (Florida yew)—a number of arthropods are apparently confined to the ravines.

Although the Apalachicola River and its swamps and floodplain have apparently been the principal highway for the ingress of continental forms, a smaller component of the northern element has moved in along the Atlantic coastal plain, and still another has entered through the western portion of the Panhandle.

It is of interest to note that in the Okefenokee region of eastern Florida, and in the Pensacola area, several continental species reach the extremes of their ranges, though the same areas are of little importance as terminal points in the distribution of Florida endemics.

The Antillean or West Indian fauna is represented in Florida by a small group of species, all but one of which are confined to extreme southern Florida or to the Keys. In view of the striking West Indian character of the flora of South Florida, the paucity of tropical reptiles and amphibians is rather surprising.

It is possible that the crocodile may be Florida's only indigenous Antillean species. It has inhabited Florida since Tertiary time; the present species, however, is Central American, having reached Florida by way of the Antilles or by circumnavigating the Gulf in times of milder climates. Of the six other species only the reef gecko, *Spherodactylus notatus,* shows strong indications of natural introduction. Its relatively general distribution on the Keys and in the tip of the peninsula, and its ecological tolerance and habit of nosing about in wave-washed wrack-piles, may indicate and account for its dispersal by hurricane and Gulf Stream. *H. septentrionalis, E. ricordii, G. fuscus,* and *S. cinereus* have all the earmarks of banana, tobacco bale, or lumber stowaways. But inquiries of the most patriarchal "conchs" I have been able to locate have convinced me that all three species lived in Key West long before they were ever discovered by herpetologists.

The hypothesis of a Tertiary land bridge connecting Florida and the Antilles, advanced by some biogeographers and rejected by noted paleogeographer Charles Schuchert, receives no support from herpetological evidence.[7]

The derivation of the endemic components of the land biota of Florida can be understood only in terms of the geological history of the state. The salient events, insofar as they bear upon the problem of endemism in Florida, are as follows:

1. The persistence of land in central Florida, in the form of large islands or a group of keys, at least since the beginning of the Pleistocene and probably since Pliocene times.

2. The bridging of the gap to the mainland on one or more occasions (perhaps first in the Pliocene), followed by renewed insular isolation.

3. Final establishment of peninsular conditions during the Pleistocene.

4. More or less extensive marginal submergences in late Pleistocene, reducing much of the eastern margin of the peninsula to a coastal archipelago.

5. Persistence of a saltwater barrier between Florida and the West Indies at least since pre-Cenozoic times and certainly throughout the period of derivation of the modern biota of the state.

At this time twelve species of reptiles and amphibians are considered endemic to Florida. From the standpoint of probable origin, these endemics seem to fall into two general groups: (1) those derived in situ, from living or extinct or subsequently modified ancestral stocks, either by isolation on a Pliocene island or islands (or, as I believe less likely, on Pleistocene islands), or else by ecesic isolation; and (2) those which represent the remnant of a once widespread pre-Pleistocene stock.

From all I can gather, the geological evidence as to the duration and even the existence of Tertiary islands in Florida is none too conclusive, and certain geologists even maintain that complete submergence of the peninsula occurred well along in the Pleistocene.

Biogeographical evidence, on the other hand, strongly indicates the existence of Tertiary land in Florida and the persistence of such islands through a very long period of time—almost certainly from the Pliocene, and possibly from late Miocene. If fairly large islands have been in existence for such prolonged periods, we may expect to find that certain organisms representing the old island biota that have narrow limits of ecologic tolerance or poor powers of dispersal have clung fairly closely to the outlines of the ancient island or group of keys on the surface of the peninsula while the hardier or less specialized or more vagile forms have become more widely disseminated.

Striking confirmation of these expectations is afforded by a remarkable paper by E. P. St. John on the distribution of ferns in Florida.[8] After several years' assiduous fern-collecting in central Florida, St. John began to notice that a number of small and very delicate endemic species were confined to a limited area in the northwestern part of the peninsula. Discussing his findings with Professor Schuchert, St. John learned that the boundaries of the region to which these rare ferns are confined correspond closely to those of the eastern part of the so-called Ocala Island—the part of the peninsula exposed in the Oligocene. Moreover, he found that "a considerable number of tropical species are represented in the United States only by their presence in this region, which is not tropical," and that "nearly one-half of the species of tropical ferns that are found in this central Florida region and in the West Indies are not found in the southern Florida region which lies directly between and where the climate and flora are tropical." The rarest and most delicate of the species are known from few and widely scattered stations, where they occur in such protected sites as natural wells and the mouths of limestone caves. They show every indication of being relict species and are almost certainly survivors from a once large and luxuriant flora of tropical origin. The characteristics and distribution of these ferns constitute strong evidence for their great antiquity in Florida, dating to a time before the cooler climates of the Pleistocene.

Unaware of the evidence of complete submergence in the Miocene, St. John attempted to explain the occurrence of the fern relicts by supposing that the Ocala Island of the Oligocene had persisted to the present. T. H. Hubbell and Sidney Stubbs note that it is only necessary to modify this hypothesis slightly to bring it into accord with more recent interpretations of the history of Florida.[9] During the Caloosahatchee epoch of the Pliocene, the warm waters of the Gulf Stream flowed across the submerged Florida Plateau, as evidenced by the tropical character of the marine fossils. The Pliocene island or islands of this epoch must have had a tropical climate and a rich fern flora introduced by windblown spores from the West Indies. With cooling of the climate during the Pleistocene the tropical species died out everywhere except in such protected situations as they occupy today.

The present coincidence of their distribution with the outlines of the old Ocala Island may probably be explained by the fact that caverns, sinks, and similar refuges were most numerous in this area, where soft and porous limestones are exposed at the surface.

Although the endemic reptiles and amphibians show no such circumstantial peculiarities as those exhibited by the ferns (and by other plant groups and many insects), there do occur certain distributional phenomena which seem difficult to explain as other than the result of insulation.

Most of the central Florida reptile endemics are burrowing forms and are found only in high pine and rosemary scrub, where the soil is deep, loose sand. Further, rosemary scrub is quite obviously the first timber stage in the succession following cessation of activity in coastal dunes, and it is apparently always replaced by high pine. Thus the present optimum habitats of those forms that are most evidently the remnant of a former island fauna, and that seem the most capable of surviving advances of the sea, which may have inundated all land except the high, dry hills and dunes, are habitats that very likely existed on those very prominences.

I am convinced that temperature plays little part in modifying the distribution of these forms in Florida. Any animal able to tolerate the astonishing extremes and diurnal fluctuations of temperature that obtain in the scrub would very likely occupy all of Florida if temperature were the limiting factor.

The argument that soil type alone may be responsible for the existence of the midpeninsular endemics—that they are the residual fragments of a formerly extensive burrowing fauna—may be met by pointing out that not one of them inhabits all the deep-sand areas within the state; the combined ranges of the Florida sand skink (*Neoseps*), the short-tailed snake (*Stilosoma*), and the brown red-tailed skink (*Eumeces onocrepis*) do not include more than two-thirds of the available areas of Norfolk, St. Lucie, and Lakewood soils. (The same evidence makes it seem very unlikely that the skinks were produced by ecesic isolation from old or modern peninsular stocks.) Moreover, the Florida scrub lizard (*Sceloperus woodi*) is not a burrower, and yet it is completely confined to the earliest stages in the dune succession—rosemary scrub or treeless dunes. Its distribution is discontinuous, there being three principal colonies: a narrow East Coast strip; a middle peninsular area (the Ocala National Forest and outlying scrubs—all probably fossil dune terrain); and the Collier County scrubs on the lower west coast. The closest relative of the scrub lizard in Florida, and perhaps anywhere, is the fence lizard (*undulatus*), which abounds in high pine but does not enter the scrub. I have walked miles along the transition zone between the two associations and have seen dozens of fence lizards on the pine side and dozens of scrub lizards on the scrub side, all within fifty feet of the transition zone, and I have never seen a single individual on the wrong side. It is conceivable that the relationship here is one of violent competition of some kind. The fact that the scrub lizard is restricted to what is undoubtedly the most rigorous habitat in Florida might lead one to conclude that it has its back against the wall, though this of course does not necessarily follow, as it is very abundant. Possibly the environment provides something that the scrub lizard requires and that the fence lizard cannot tolerate. It is perhaps significant that in other parts of the state there are patches of rosemary scrub that the fence lizard approaches but does not enter and in which the scrub lizard does not occur at all. Whatever the relationship between the two at present, I am inclined to regard the *woodi* as an island dune derivative of some old *Sceloperus* that may or may not have been *undulatus* but which probably was the forbear of both.

The herpetological fauna of Key West, the terminal point in the continuous portion of the archipelago of keys that extends southwestward from the southern tip of the peninsula, consists of some thirty-two forms. Of these, nineteen are continental or Floridian species or subspecies that, by one means or another, have reached the island to mingle with the West Indian element discussed above. What the fauna of Key West was before the advent of shipping and railroad and highway connections is a matter of conjecture. In all probability the majority of the species have been introduced by human agency. It is of interest to note, however, that this group is largely composed of the same forms that make up the faunae of the coastal islands of the peninsula. It thus seems reasonable to suppose that they must possess certain peculiar qualifications for fortuitous dispersal and for tolerating the severities of insular life.

Photographs from the Life and Times of Archie Carr

Captions by Marjorie Harris Carr

A young friend of Archie's, May 13, 1921. James Henry Lee, Billy's Island, Georgia. (Photo by Archie Carr)

Archie at age ten with turtles, 1919. He is in a friend's backyard near his home in Fort Worth, Texas. (Photo by William Stitt)

Archie at age sixteen with two friends. They are sailing in the *Capermacon* on the Wilmington River near Savannah, Georgia. Archie is at left.

University of Florida biology majors at Paynes Prairie, 1930s. The students are rolling hyacinths up onto the bank at Haille's Siding on the north shore of the prairie. Scratching through the rolled-up plants was a favorite method of collecting small aquatic vertebrates and invertebrates. *From left:* Frank Young, Archie, and John Kilby.

Archie, a senior at the University of Florida, 1932. He is prepared for a collecting trip on Newnans Lake, near Gainesville.

Touring car equipped for collecting in Florida in the 1920s and 1930s. In those days Florida was as much a mecca for field zoologists as the West Indies and Central America are today. Dr. Albert Hazen Wright and his wife, Anna Allen Wright, specialists in amphibians at Cornell University, drove to Florida in this car many times. Archie and other graduate students were their enthusiastic field assistants, and Dr. Wright was probably the first conservation activist Archie knew. During the 1920s and 1930s, when plans to dig the Cross-Florida Ship Canal (also called the Ocklawaha–Withlacoochee Canal) were coming to a head, Wright published "The Atlantic-Gulf or Florida Ship Canal," a small booklet that pointed out the environmental hazards posed by the canal. Work on the canal stopped after only six months, mainly because of lack of money, but Wright's report played a significant role in the project's suspension. Archie, who contributed to Wright's early publication, was still working in 1987 to save the Ocklawaha from the ravages of the Cross-Florida *Barge* Canal.

Thomas Barbour, J. Speed Rogers, and Archie, Gainesville, 1942. Dr. Barbour, known as T. B., longtime director of the Museum of Comparative Zoology at Harvard University, had an enormous zest for life and an intense appreciation of the natural wonders of Florida. He would spend springs in Gainesville, and we spent summers in Cambridge for several years. He was a grand raconteur and larger than life in every dimension—a wonderful man. Speed was head of the department of biology and geology at the University of Florida between 1922 and 1946. He and his close friend and colleague Dr. Theodore H. Hubbell were chief architects and practitioners of the ecological approach to the study of animals. They had incalculable influence on Archie and numerous graduate students, who spread the word throughout the nation about this approach. Archie is holding our dachshund Zep.

Archie taking a break while looking for crocodiles in Madeira Bay, Florida, 1940. At that time little was known about the status of the crocodile in Florida. Today crocodiles are much better understood and are fairly safe in the seven-thousand-acre Crocodile Lake National Wildlife Refuge.

Opposite, field trip to the River Styx, February 1966. For more than twenty years Archie taught ecology at the University of Florida. Weekly field trips to different Florida landscapes were the highlight of the course. Top, Archie with students on the floodplain of the river. Center, Archie dredging for small aquatic animals in the roots of stranded hyacinths. Bottom, the class examining the booty; David Ehrenfeld is facing the camera. (Photos by William M. Partington)

Archie with student on Snake Key, 1968. The key is so named because it is home for a large population of cottonmouth moccasins that live in a mutually beneficial association with the nesting water birds. (Photo by Jo Conner)

Archie and his son Stephen watching fishermen landing sturgeon on the Suwannee River, 1976. In 1985 the state of Florida designated the sturgeon as a "species of special concern" and prevented any further harvest. Starting in 1975 Archie collected information about sturgeon in the Suwannee and Apalachicola rivers. All of his sons took part in this project, and since 1985 Stephen has been tagging and following the movements of sturgeon in the Suwannee with amazing success.

The whole family lined up for a Christmas picture, Micanopy, 1954. From left: Tom, Archie, Mimi, Stephen, Chuck, Marjorie, and David.

Wewa Pond, 1968. Wigeons, gadwalls, blue-winged teal, ring-neck ducks, and green-winged teal came by the thousands that year to feast on—of course—duckweed. (Photo by Archie Carr)

Archie with Ben, 1970s. Ben is examining a young green turtle—not an inhabitant of Wewa Pond.

Wewa's alligator, 1980. The gator is guarding her nest, the grassy mound to the left. She has shared Wewa Pond with us for more than forty years. (Photo by Archie Carr)

Jasper, an alligator snapping turtle, 1964. Jasper was given to us by a fisherman from Jasper, Florida. After he lived for a few months in a small pool, we placed Jasper in Wewa Pond. He became quite tame and would come when called. Here David is offering him a piece of catfish. (Photo by Archie Carr)

Archie at the visitors center of Paynes Prairie State Park, 1976. Paynes Prairie is fifty square miles of level plain formed by the collapse of the limestone bedrock. It drains into a large sinkhole, Alachua Sink, which used to clog up from time to time, causing the prairie to become a shallow lake. When dry, the prairie is a superb natural pasture, and the Seminole Indians maintained large herds of cows and horses there in the late eighteenth century. (Photo by Herb Press)

Archie in the Fahkahatchee Strand in southwestern Florida, 1972. The Fahkahatchee, a strip of wilderness some twenty-five miles long and seven miles wide, includes samples of almost every landscape found in the big cypress swamp. Here, air plants cover the branches of the pond cypress and other host trees.

Florida Vignettes

There are still remnants of the old wild Florida. There is always something. Anytime. Day or night, cold or warm, in the rain or shining sun you can find bits of the old wild left around, if you can only get away from your fellow man for a spell. Living as I do there is always something.

The Spider

I found the spider on a low bay leaf over the edge of a lake in Umatilla. I was down there one afternoon long ago hunting spiders along the shore for H. K. Wallace, who was then a fellow student working toward a Ph.D. in spider biology at the University of Florida. My parents lived by the lake, and so did the spiders Wallace wanted; and I was down there peering among the wax myrtles, buttonbushes, and sprouts of silver bay at the edge of the tea-red June water for diving spiders when I came upon the spider I spoke of. It was a kind I didn't know—a jumping spider, I suppose, but with a longer body than they usually have. Anyway, I was about to pass on down the shore when I suddenly noticed that this spider was creeping up on a naked measuring worm that was resting quietly at the edge of a leaf. I stopped to see the outcome, sure it would be the usual faultless charge, with the poor worm ending up a thin sack with all its juices gone. But this time something went wrong. The spider jumped, and the worm dropped off the edge and drew up swinging by a thread, two feet below the leaf and a foot above the water. You have seen measuring worms do that, I am sure. I don't know whether they anchor their lifeline at the instant of crisis or keep it fast all the time to what they are walking on, for safety's sake. In any case, when the spider missed his charge and the worm fell free, it was not just to be snapped up by the stumpknockers in the dark lake water below.

"The Spider" from Archie Carr, *Ulendo: Travels of a Naturalist In and Out of Africa* (New York: Knopf, 1964; Gainesville: University Press of Florida, 1993).

I looked closely to see how the spider would take the reverse, somehow feeling he would not simply go away or settle down in melancholy. I was right, too. The spider clearly perceived that the loss was not complete. He walked rapidly back and forth for a moment, waving his pedipalps thoughtfully about, then he reached down with a front leg and felt along below the edge of the leaf and quickly found the strand that held the worm up. I am sure he couldn't see the worm itself. The spider was on top of the leaf, you understand, and his eyes were all on top of him; there was no way for him to see a thing two feet down toward the water. But in spite of this, the instant his foot touched the taut silk he began to pull it in. It was a worm-silk strand, remember, for all I know spun of wholly different stuff from spider threads. But the spider grabbed it as if he had reeled it out of his own spinnerets and began to haul it in hand over hand, passing it back to the pairs of legs behind. The second legs took the loops with deft ease and handed them back to the third, and they in their turn gave them back to the fourth legs and then reached forward again for the new loops coming. It was skilled work but wholly orthodox for a spider, and in only a short while the worm had been drawn up almost to the edge of the leaf. Still not looking down, the spider suddenly stopped pulling, moved out to the edge, and shifted about for a stance that would let him get at the prey. Feeling around for what to do, he ran one of his feet down the shortened line. It touched the worm lightly, and instantly the worm dropped down a foot and bounced to a swinging stop in the safe space between the leaf and the water.

Again the spider began to take in line. He did it with a little less sure stealth this time, with what I took for a tiny taint of worry. Again he was able to raise the worm to within millimeters of the platform, and this time, too, when he groped for a position from which to snatch or stab or do whatever he had in mind, the worm took fright and fell free and swung there slowly above the water.

Up to then I had only marveled at the sense of the spider—at his knowing, without seeing, that his quarry was hanging below the leaf, and his seeing so quickly how to fish it up. But when the spider again began to pull in line, I could see the incident taking a different turn. I began to understand that the shrewdness of the spider was not sufficient to show him what to do with the worm thread he was drawing in. There seemed, in fact, to be in him a sort of gradient of shrewdness, going back from his front end, which had thought up the project, through the substations in charge of the four pairs of legs. As the line was passed along back, it was the fourth pair of legs that had to decide what to do with it. Why not just drop it, one might say, and perhaps that is what the front end thought would be done.

By the time he had begun to draw up the worm after its third drop, grave trouble was mounting. The second legs still shared the optimism of the first. They took the thread with confidence and passed it to the third. But the third legs quite clearly showed diminished verve in handling it; and on back at the fourth pair the insight anteriorly was beginning to build disaster. Already the fourth legs had received more thread than they knew what to do with. The stuff was too light and flimsy to be thrown off the edge, or even to lie still where the frenzied combing of the hind feet tried to drop it. Instead, it piled and tangled and clung to whiskers and spines of legs and feet—and still it came looping back in more air-thin bights and hitches. Before long, the last legs had lost all semblance of ability to cope with the clinging stuff, and their uneasiness began to flood forward in the spider. The third legs were stirred to reach back and try to help with the harassing tangle. But the silk kept coming, and because the last legs were unable to take in any more, the third were soon in as bad a state as they, and the second pair started reaching back to help in the growing crisis.

By then the worm had been pulled up almost to the edge of the leaf again. When its hind end showed at the rim, the spider launched a short jump, but, with most of his legs trussed up with worm silk, he missed. The worm let go and fell again and dangled as before. Still clinging stubbornly to his insight, the spider—the front part of the spider—fell to snatching in the worm-string in a way that bordered on hysteria. He dragged it in over the edge with unnecessary vigor. He handed it back in heedless abandon, showing no concern for the chaos back there, where six of his legs had lost all their initial enthusiasm, all thought of group action, and were only scraping miserably at themselves, helpless in the toils of parochial anxiety.

I give you my word, it was a bad business. There was this poor spider with his front end still filled with insight, hauling away at the hanging worm, while his afterparts were taken hopelessly aback by the versatility in the front, all their reflexes shot to pieces. And still the forefeet hauled in line, more frenziedly, even, than before; dragging it up, pushing it to the rear, and leaving it lying there untended by even the second legs, which by now were distraught with the gravity of their own predicament.

I don't pretend to know how realization comes to a spider, how a dawning doubt that starts at the back end moves forward through the subdivisions of his nervous system. But in this case I'm telling of, when the tangle, and the apprehension, had spread forward to the second legs, and they had begun to help the third and then to turn back and claw at their own snarled quota of the worm's thread, you could see desperation mounting in the spider as a whole. One moment he was pulling frantically. Then all at once you could feel him sensing back through the grades of panic behind—and then suddenly he stopped his pulling. With one front foot he reached back as if to assess the disorder, testing with a single nervous touch the tangle about each hinder pair of legs. The appraisal done, the calamity sure in his mind, he stood there for a moment and waved his pedipalps. Then sadly but quickly he drew the thread up to his jaws and bit it off.

The worm fell into the black water of the lake with a plink too tiny to be heard. I picked him up before the stumpknockers could get there and put him on another leaf. As I went away the spider had turned to the work of chewing and clawing free from the silken bonds the worm had left him.

This is a peculiar sort of tale, I realize. But I hope the main moral is clear. What the spider did of course required no insight at all. Spiders live by pulling strings. They weave and trap prey with string webs, and they throw nets and anchored bolas at bugs. They hang up egg sacks and swing from lines or walk on them, and they truss up victims and haul them here and there. They pay out cable, rig guys, take in slack, and eat up spare line to save it. Scare a spider and he does what the caterpillar did—he drops and hangs by a quickly fastened safety thread. Knowing about strings is built into spiders. It surprised me to come upon a spider that could think so quickly across the gap between pulling his own string and pulling that of another creature. But the gap was pretty narrow, really, and the need for insight was little or none. The spider just made a slight adjustment in fixed behavior his ancestors had long before acquired.

Gopher Tortoise

There is a gopher in the pasture by our house. His burrow is in the middle of the plot of close-cropped Bahia grass. Each suitable day the gopher goes out to graze—or browse, I suppose you could call it, though I am not sure. Anyway, he goes out to eat cactus. Our pasture is badly kept, and patches of downtrodden opuntia are scattered about in it. The gopher seems never to eat any Bahia grass, only opuntia, so his daily grazing regimen takes him from one of the cactus patches to the another looking hopefully for whatever stage or state of cactus it is that pleases him.

This gopher and I have a game that we play. At least it is a game with me. For the gopher it is no doubt miserable harassment. He is very fast and very self-conscious, living out in the open pasture as he does. What I do is try to catch him before he can get to his hole. The hole is out fifty feet or so from the nearest concealing brush along the fence. So every time I see him grazing out some distance from his hole I sneak up to the fence and crawl under it, then rise and start running, not toward the gopher but toward his hole, because it is a sure thing he will run for it too. I have done this half a dozen times, and never have I reached the hole ahead of the gopher. One time the race began when he was at least thirty feet out on the far side of the hole. We then ran the race straight toward each other. We hit the hole together, me with one outstretched hand like a runner barely safe at home plate, the gopher in a precipitous dive that my hand could not stop.

If there is one land animal that more than any other epitomizes the appeal of the old wild Florida, it is the gopher tortoise. When people were few in Florida, gophers abounded, but those days are gone, and so too are gophers from most of their range. Their decline is one of the melancholy repercussions of human inroads into the natural landscapes of the state. The diminishing colonies that remain deserve scrupulous protection—both the animals themselves and their burrows, which are ecologically important as habitat for a long list of regular and occasional guests.

A worrisome aspect of the decline of the gopher is the lack of any way to measure it. Nobody ever took a census of the gopher population in the old days, so there is no basis for comparing what is left to what once was here. As a result, the gophers are dwindling without the fanfare due such an aesthetic and ecologic loss.

There are even some people who assure you that gophers are more numerous than ever. They tell you people used to eat them all the time and don't any more. So there's bound to be more gophers, they say.

Well, for one thing, people do eat gophers nowadays. Regrettably, they are good eating, and rural folk of all creeds and colors eat them at every opportunity. But the main dimming of hope for keeping gophers in the landscape is the changing of the landscape itself.[1]

Glowworms and Fireflies

I saw the glowworms when I went down in the dark to put out scratch-feed for the teal. The whole south end of Wewa Pond, where it wets the edge of the yard, was bordered with a belt of little lights. They may have extended on out into the wooded rest of the shore, but if they did, the weeds and bushes hid them.

I stood in the band of my flashlight and tried to figure out what the glowworms were doing. I had seen lots of glowworms in my time but always as isolated single glows, in twos or threes or other feeble numbers— never in anything you would want to call a crowd.

One usual excuse for such crowding together by bugs is to breed. The lightning end of their parents, the fireflies, is, in some cases at least, a part of their mating apparatus. But glowworms are larvae and too young for that. So I decided that the glowworms must work a zone of pleasant country or that they liked the damp there or that they were eating something that had found its own aggregation in the zone. Glowworms are carnivorous and have been found eating snails. So I got down on my knees by the light of the flashlight, searched carefully among them for snails, and found none at all, nor anything else that seemed edible.

After a while I spread the scratch-feed for the ducks along the edge of the water and went back up to the house, leaving unexplained the presence of the host of worms and their unaccountable glowing.

The decorative flashing of fireflies is mainly a breeding activity. The rhythm of flashes, phrases of flashes, and pauses between flashes serve to attract the sexes and to identify both the species and the sex of the bug doing the flashing. In some species at least, the pattern inscribed across the dark by the lit-up firefly is a part of the courtship and recognition play.

When courtship time comes, the male firefly takes wing and goes into the light-dance of his race. The female answers with a flashed signal, and the male goes to her. The phrasing of the flashes is as peculiar to each species as its body form or coloration or any specific attribute. Not even the most irresponsible female will answer the signal of a male not her own species.

This peculiarity of fireflies is useful to the entomologists who study them. The females of some are hard to collect. They are secretive, and the only lighting-up they do is a flashed response to the signal of the male. Once the courtship signal of a species is known, however, you only have to go out with a penlight or small flashlight and faithfully go through the series of flashes, pauses, and movements of the species you are interested in. If you do it right the female will give an answering flash.

Dr. James Lloyd, a zoologist at the University of Florida, has shown that entomologists are not the only ones who take advantage of fireflies this way.[2] Other fireflies do, too. Lloyd has found that females of the carnivorous genus *Photuris* mimic the reply signals of females of *Photinus,* another kind of firefly. The *Photuris* females perch about on twigs in places frequented by *Photinus,* and when a male explores the dark with his amorous signal, he is answered so reassuringly that he flies over, full of hope. When he reaches the flashing female he finds not a mate but a burly voracious firefly three times his size who quickly grabs and devours him.

Frog Songs

One of the less well advertised virtues of Florida is its distinction in respect to frogs. There may be more kinds of frogs per acre in other parts of the world. Brazil is full of frogs, for example. But no other state has so many kinds or such volume of individual frogs; and certainly none has the varied frog song that Florida has on appropriate nights when the birds have gone to roost.

The songs of some of the Florida frogs are similar, but more often they are at least as distinct as the looks of the creatures, and sometimes a lot more so. Take the tree frogs, for instance. There are eight kinds of them in Florida; they are all relatively small, moist, long-legged little frogs with flat pads at the ends of their toes. And some of the kinds are so much alike that they must be looked at sharply to be told apart. But each can be instantly identified if one quick snatch of its main breeding song is heard.

I once wrote a key to the songs of the Florida frogs, and a user with a good ear and flexible imaginative outlook could actually key out—that is, identify—frog songs with it.[3] My key was not in the vein of modern frog song analyses in which the singing is taped and a sonogram made to show its every nuance of pitch and volume. Mine was a more subjective key, independent of electronics and dependent on a certain amount of listening ability—and, I suppose, on the ability of the listener to hear as I do. So the key never got wide use as a practical way to identify frogs from their songs; but reading it does give one some idea of the sonic diversity that frogs lend to a Florida night. Very briefly, tree frog songs may be described as follows:

1. *Hyla gratiosa:* bonk

2. *H. cinerea:* erp

3. *H. squirella:* ack

4. *H. femoralis:* tiki-tiki-tiki-tiki

5. *H. crucifer:* whistle, low notes, one up

6. *H. avivoca:* whistle, one note rapidly repeated

7. *H. versicolor:* whistle, short trill

8. *H. sepentrionalis:* harsh aark

Once in a while we get a letter addressed to the zoology department of the University of Florida from some citizen, usually one newly arrived in the state, asking advice on how to kill frogs. Some of these letters are from people agitated over the spread of the great toad *Bufo marinus*, which is now firmly established in southern Florida. Needless to say, that kind of worry is baseless because a toad can't hurt people unless they bite it, and few people bite toads; those that do can be considered expendable.

Anyway, most of the letters I have seen were not about toads at all but were complaints about the hideous din of frogs on summer nights when the writer wished to converse with friends or sleep or watch television. I can't say I actually sympathize with those people. There is more intellectual reward in a rousing frog chorus than in much of television, but it is good to hear and be heard by your family at times, and sleep is to be cherished always. So even though I consider the attitude of the frog-letter writers extreme, at least I understand what their problem is and admit it can be a bit of a nuisance.

The main offenders in the particular encroachment of wild nature upon man's preserve is probably the green tree frog *Hyla cinerea*, though if tallies could be made, the common toad would not be far behind. The two reasons—one for each case—that these creatures disturb man more than most are that the toad lives in greatest abundance about houses, so it aggregates more than most in swimming pools, yard fish ponds, and other domestic water at breeding time; and that the green tree frog breeds in permanent, well-established bodies of water instead of in floodwater like most other frogs, and so it comes in contact with people who have homes on the waterfront.

This morning the belated spring frogs at last came out. They usually show up with the first warm rain of March or in April at the latest. So far this year we had a strange silence in the woods around our house. Out in the pond the green frogs sang from the reed stems, cricket frogs sang on the edges and floating islands of the pond, and sulphur-bellied frogs grunted at each other from the few gator-proof retreats set about the pond. But that was all we had of frog song till now. Last night it rained a frog-choker rain, and the neighbor's land came sluicing down into the trough of our woods road as almost never before. Now evidently the wet has soaked down comfortingly through the leaf mold and into the soil where all the burrowing creatures pass their quiet season.

I woke up before dawn and heard the massed trill of toads and knew that finally the little frogs had come or would soon be coming. And later in the morning when my wife and I walked down through the woods to

Smokey Hollow, *Hyla squirella* was acking from the dripping trees and the Morse-code call of the pine tree frog—*tiki-tiki-tiki-tiki*—was sounding in a scattering of higher trees. So barring the untoward barging in of outside cold dry air there will be frog choruses tonight. I hope there are, and I hope they allay my unease over the suspected silent spring of the little frogs.

Tails of Lizards

It was the old lizard coveting the wren's worm that got me thinking about skinks a while ago. I was having breakfast at the time. Looking through the glass door beside our terrace I saw a big lizard come out from under the sofa where it lives and set out on its foraging patrol along the baseboards. The lizard was a skink, a mature female *Eumeces laticeps,* around nine inches long with smooth, lustrous scales; she was blue-black in ground color with a light stripe along each side. She had grown up on the porch, eluding coral snakes in the border plants, becoming accustomed to the passage of people, and even tolerating the attentions of extroverted young dogs without much uneasiness.

On summer nights, insects come charging up from Wewa Pond to the lights of our house and either crowd the lighted windows, to the delectation of the tree frogs, or bash their brains out on the terrace walls. It is this steady fall of incapacitated bugs that the skink finds attractive.

On this morning a wren was hopping back and forth beneath a row of hanging plant baskets. A pair of Carolina wrens builds a nest in one of these baskets every year, stoically ignoring my wife's occasional watering of the plant in it. The wrens were introducing their newly hatched offspring to insect food, and apparently their appetites were hard to awaken. As each parent arrived with an offering it had to wait its turn to enter the nest. The delay seemed stressful, and the frustrated parent passed the time jumping nervously back and forth on the floor and looking anxiously up to the nest. It was during one of these times, when one of the wrens was doing its impatient little dance with a big naked writhing caterpillar in its beak, that the old skink appeared around a corner of the wall.

Seeing the agitated bird six feet away, the skink cocked her head, tasted the air with her snakelike tongue, wriggled her tail slowly, then made a quick sally out to within a foot of the dancing wren. With surprising un-

From *Animal Kingdom* 88, no. 3 (1985).

concern, the bird just hopped aside a few inches. The skink made another approach; again the wren reacted with a quick but casual hop. Four times the skink repeated the fruitless overture, and there was no escaping the conclusion that her aim was to take the worm from the wren. It seemed strange that she should have thought this possible; but then, so was it strange that the wren showed so little concern about the encounter. Shortly after the fourth approach the wren at the nest departed, and the one on the floor flew up to make its delivery.

A habit of skinks that has evoked much theorizing is the way they move their tails when foraging. A hunting skink moves forward a foot or so, pauses and tests the air with its protrusible tongue, then, in a seemingly purposeful way, undulates its long, tapering tail in graceful sideways serpentine waves. For decades people have been trying to figure out why skinks do this. One theory is that the maneuver lures prey. Another is that it deludes predators, distracting attention from the vulnerable head and midsection of the lizard and focusing it on a disposable part.

And the tails are indeed disposable. Like those of geckos, glass lizards, and several other species, skink tails break off easily under stress—even psychological stress. When this happens, the tail doesn't just lie around dejectedly, doing nothing. It starts jumping all over the place and keeps this up for some time.

Yet another striking feature of skink tails is their electric-blue color. I don't know how widespread brightly colored tails are in the family as a whole, but they occur in the young of several species of *Eumeces,* and in the females they linger into maturity. Thus, when the tail of a blue-tailed skink cuts loose and goes into its frantic break-dance routine, it becomes by far the most conspicuous object in the vicinity—one ideally designed to cover the retreat of the erstwhile bearer.

It seems to me, however, that when a skink makes its customary pauses during foraging it might be better off to just lie still. The tail movements seem likely to attract the attention not just of attacking enemies but also of distant predators that otherwise might not notice the lizard. To be sure, once an attack is launched, the wriggling tail might draw attention to the rear; but without the conspicuous undulations maybe the whole skink would escape notice.

When you try to pin down the adaptive value of the vivid blue of the tail the problem grows even more complicated. Three possible explanations suggest themselves. The color may simply reinforce the supposed distracting effect of the undulations of the tail while it is still on the skink and the acrobatics when it is broken off. Or perhaps both the color and the way the skink waves its tail have evolved as means of attracting prey rather than distracting enemies.

Brightly colored tail tips occur in several kinds of snakes who also wave the tail tip around in a purposeful, stereotyped way, and observations have shown that the ruse does attract prey. The third possibility is that the colored tails are aposematic—conspicuous and serving to warn—like the color patterns of coral snakes and monarch butterflies. Creatures that are noxious or poisonous when eaten or that deliver venomous bites or stings often advertise the trait by means of bright colors or vivid patterns. And the blue tail of a skink looks for all the world like a sign that the bearer is either venomous or toxic. So the question is, are skinks toxic?

As a matter of fact, all the way from North Carolina to Panama people unversed in herpetology call one kind of skink or another *scorpion* and consider them to be venomous. I have no idea how the notion spread abroad, but I have heard it from Georgia Crackers and Mexican Indians and Creoles and Miskitos of the Caribbean coast.

Skinks are certainly not venomous. Whether they are poisonous if eaten, however, is not so clear. Many natives of the southeastern United States believe that skinks poison cats that eat them. This has never been proved, but it is not just a folk belief like the lingering notion that skinks whistle in flatwoods ponds on rainy nights. A cat malady involving gastric symptoms and paralysis is widely known in Florida as lizard poisoning; and I know a lot of sensible people who are convinced that the symptoms appear when cats eat skinks. Or the tails of skinks. Some think the poisonous principle resides in the tail.

The numbers of U.S. citizens who believe that the bite of a skink is venomous have diminished, but the idea lives on that eating *Eumeces nauseates* paralyzes or otherwise damages cats.

I once canvassed well-educated, serious-minded friends on the subject and found them either open-minded or firmly convinced that lizard poisoning is fact. Many of these people have cats, and some have taken their cats to the vet when they appear to be suffering from what the owner considers lizard poisoning. None of my informants claimed to have watched his or her cat eat a skink and then show distress, but some had found the lacerated, usually tailless body of a skink after the cat got sick.

When a colleague came to school distressed because his cat had eaten a skink and nearly died and was left holding her head sideways all the time, I decided to find out what veterinarians thought about the matter.

The inquiry was frustrating. I talked with eight vets in four towns, two states, and two colleges of veterinary medicine. They all knew the so-called feline-lizard-poisoning syndrome very well but considered it a folk invention. When I asked why the idea has hung on so long, they said people just grew up with it and couldn't get rid of it. When I asked how they, the vets, had come to discredit the belief, they said it had never been scientifically proved. When I asked what experiments had been made to get the straight of the matter, the answer was none. One doctor had ground up some lizards and fed them to cats, and no symptoms resulted; but he had used anoles, not skinks, so the test was irrelevant.

If Florida skinks are really poisonous, then you just about have to think of the blue tails as a warning signal; and this, of course, would in no way detract from the utility of the color in drawing attention to the dancing tail after it has broken off.

Back in the days before the armadillos came, little brown ground skinks, *Scincella lateralis,* abounded in the hammocks of North Florida. When I was a graduate student I looked into a lot of their stomachs to see what they ate. In four that had recently lost their tails, I found end sections of *Scincella* tails, and in a paper on Florida herpetology I suggested that the skinks had bitten off their own tails. In support of that notion I adduced the following fey little observation: "George Van Hyning and I once watched a captive *Eumeces laticeps* regard its tail for a moment, wriggle it tentatively, seize it in its jaws, break off the terminal inch-and-a-half, and swallow it avidly. The inference of autophagy in *Leiolopisma* [*Scincella*] does not appear to me to be unfounded."[1]

There was, of course, no proof that the tails the bobtailed ground skinks had eaten were their own and not those of opponents in some fight. But the fact remains that our pet *laticeps* did find her wriggling blue tail tempting and did bite it off and swallow it. And this must in some way bear on the questions discussed above, though I'm not sure exactly how.

*A*lligator *C*ountry

Almost anyone who is asked to say what animal or vegetable most clearly epitomizes the color and spirit of the primeval Southeast would say the alligator. In reckoning the natural assets that first brought Florida to the attention of the outside world and later made some of it the most glamorous real estate on earth, the role of the alligator would be hard to overestimate. The alligator is not, of course, an exclusively Floridian production, but when most people think of Florida they think of alligators. And because Florida is growing faster than the rest of the Southeast, and because it had more alligators to start with, it is in Florida that the decline in alligator populations seems most dramatic.[1]

The American alligator has been quietly giving ground for a century. In pioneer days it was an active fear that turned the muskets of the scattered settlers against the gator. Although some fear survives even today, the real troubles of the alligator are the ruin of its habitat and the soaring prices for its hide. Only a short while ago an alligator brought the hunter less than a dollar. By 1968 the price had risen to seven and eight dollars a linear foot, and the poachers, along with the drainers, fillers, and bulkheaders of the landscape who perennially plague the state, began wiping out wild alligators at a sickening rate.

It is not the alligator alone that is facing extinction before the spread of humanity. It is the whole reduced representation of the crocodilian clan—the two dozen kinds of caimans, crocodiles, gavials, and false gavials that, besides the two surviving kinds of alligators proper, are all that remain of a once-dominant reptilian line. These are all that are left of the old archosaurs or Ruling-Reptiles, the main stem of reptile evolution that founded much of the vertebrate life of today. The ancestral alligators kept the old conservative four-footed, low-slung, mud-slogging anatomy, went in for no elaborate changes in design, and in spite of drastic decline lived on until today.

Obviously it has not been the species we call the American alligator that has lived all that time, but his family and his order. Alligators as fossils are first known from the Miocene, some thirty million years ago. Today there are two kinds of true alligator, distributed in a peculiar way. One is *Alligator mississippiensis,* of the southeastern United States. Its natural range extended from North Carolina to the Rio Grande and northward in the Mississippi Valley to Arkansas. The other species, *Alligator sinensis,* lives in China, where it is today confined to the inhospitable marshes of the Yangtze delta. Clifford Pope, the first western zoologist to see this creature in its native habitat, believes that it formerly was widely distributed in Southern China. The only reason there are any Chinese alligators left is that the Yangtze delta is such uninviting terrain that even the overcrowded Chinese have not yet settled there. The anomalous geographic separation of the two kinds of alligator was not caused by one of them wandering away from the other but by the gradual dying out of intervening stocks of once widespread ancestors.

The alligator line is thus one of the classic manifestations of life on earth. Throughout all but the tiniest fragments of its original range the alligator is disappearing. To save him as a feature of the natural landscape will take fast work.

If people have anything approaching a valid complaint against alligators nowadays it is over their unmannerly feeding habits. They are very open-minded regarding food, toward both the kind and the size of the parcel it comes in. They eat almost any kind of animal food available. The ponderous jaws combine great crushing power with a surprising precision in striking at small prey—snapping delicately at butterflies dancing over the water or at dragonflies on floating cabomba blossoms. Alligators are especially fond of marshmallows. It is a recently acquired taste, of course; but nowadays marshmallows are the thing that you take with you to a Florida nature preserve or reptile show where alligators are on display, just as peanuts are taken when elephants are visited. What attraction marshmallows hold for a gross, carnivorous reptile is hard to imagine. Maybe their soft, springy feel reminds the gator of the belly of a frog that has puffed itself up with air, as frogs do to avoid being swallowed. In any case the marshmallows are a great convenience for visitors who want to establish some personal contact with alligators along nature trails and yet who feel unable to throw them chickens, frogs, pigs, or other things that generally come to mind as alligator feed.

There are, however, good reasons to doubt the advisability of allowing the public to feed alligators anything, even in wildlife preserves. The problem is that alligators readily fall into the custom of watching out for edibles thrown down by people on banks and boardwalks. Their enthusiasm over this intercourse with humans makes one wonder why an alligator might not interpret the odd child falling into the water as some offering thrown down. It is really pretty visionary to expect a rending gang of carnivorous reptiles to distinguish between the beef bones that some people throw them and the children the people bring down to watch them eat.

But to go on with the more natural feeding habits of the animal: they are varied. Alligators are really pretty indiscriminate, opportunistic feeders. It is impossible, in fact, to say more than that whatever can be caught is eaten. An alligator living in a natural environment—that is, a place not affected by flood, drought, or famine—just goes around piecing out his diet as chance allows, seizing the unwary or ill or confused among any kind of animal that may come within reach and not eschewing corpses or carrion whenever these come to hand.

Insects are important in the diet of the young ones. Frogs are taken by alligators of all ages, and they eat fish and turtles at every opportunity. The shortnosed gar is a staple food in some localities, but fish are hard for an alligator to catch and predominate in the diet only where concentrated by low water. If alligators are there during such low-water times there can be mayhem among the fishes, and observers may go away muttering that alligators are great enemies of fish. They are not.

Alligators eat turtles, snakes, and bullfrogs. Birds of all kinds are eagerly hunted but probably not often caught when in full possession of their faculties. All the mammals with any aquatic tendencies—round-tailed muskrats, otters, rice rats, swamp rabbits, raccoons, opossums—are taken whenever they get unwary. Crayfish are important in the diet of the young in some places. Salt marsh alligators eat a lot of crabs and shrimp.

Like the Old World crocodile, alligators are specially adapted for ambushing creatures that come down to dabble or drink at the edge of the water. All the crocodilians, in fact, except perhaps confirmed piscivorous gavials, probably share the habit of lying near shore and sweeping into the water, by a sudden traplike bowing of the body, any dabbler unaware of the different looks of logs and alligators. The maneuver simultaneously knocks the prey off the bank into the water and crowds it into the gaping jaws. In this way crocodiles eat many African women who go down to the shore to wash clothes on rocks in shallow water. The American alligator is adept at the same technique but never uses it for catching women.

One of the really bizarre aspects of the personality of alligators is their reluctance to eat people. It cannot be that gators have ethical taboos about the matter; and it certainly is not merely that people are too big. Alligators eat dogs and hogs, and some people are no bigger than a dog or a hog. Anyway, gators have even been known to drag down cows. In the old days, abundant big alligators probably collaborated in their feeding—that is, fought and struggled over a victim too big for any one of them to swallow and twisted and rended it in a way that pulled it to pieces. Crocodilians have no shearing or cutting teeth and so must tear food apart if it is too big to swallow whole. Strife at feeding time, therefore, was advantageous because it let the collaborators eat animals of any size. In the few places in Africa where crocodiles and big game are still plentiful, the crocodiles use the same technique. The tendency of even the tiniest alligator to grab hold of an edible object too big to swallow and then spin in the water till a piece is torn away suggests that meals on wild hogs and bison calves were a part of the near past of the marshmallow-eating alligator we know of today. There is no reason at all to believe that alligators have a built-in aversion to big prey. Eating people, however, seems to be against some rigid rule they have.

It is not easy to figure what the cause of this reluctance may be. One line of reasoning simply suggests that for a hundred years or more alligators have been picked on by mankind, and three things keep them from eating people: they are smaller; they are fewer, and so neither compete so hard for food nor get so hungry; and they have learned to be afraid of people—that is, existing alligators big enough to eat humans have all been shot at or otherwise scared by man, and so avoid him.

But if you accept this argument it means you accept the corollary implication that in the old days, when the southeastern United States was thinly settled, alligators did eat people. It is not easy to get reliable information on that point, either. If the outlook of alligators did actually change at some time in the past—if they changed from man-eaters, or rather from indiscriminate eaters of anything they could catch, including people, to selective victualers that eat anything they can catch except people—then the change must have happened a very long time ago. It had to have been longer ago than a hundred years, and maybe much longer.

The early impressions of alligators recorded by travelers were nearly all morbid. The first published illustration of an American reptile was probably Jacques Le Moyne's picture of a gigantic alligator being mistreated by Florida aborigines (see drawing).[2] In the legend to the figure Le Moyne explained that this was the way the people had to spend a lot of their time—protecting themselves against their implacable foe, the alligator. This was a very influential picture that did much to set the tone of dread that has tainted human feelings about alligators. There is an ironical twist to this too, because though the picture was "drawn from life," it obviously misinterprets the scene it shows in a fundamental way, and in a way unfair to alligators. The Indians in the picture are all shown holding on to a huge

"Indorum Floridam provinciam inhabitantium eicones," drawn by Jacob Le Moyne, in Theodore Bry, ed., *Voyages en Virginie et en Floride* (Liège, 1591).

pole—the trunk of a good-sized tree, really—that projects from the mouth of the alligator. By the set and slant of the stances of these Indians you would say that they were pushing the pole into the alligator's mouth, as if to fend off an attack. It is the clear intention of the figure to give this impression.

Now I am about to propose an ethnozoologic correction that may lessen by a little the weight of evidence against the alligators of primordial Florida. My thought is this: maybe that artist did see an alligator with a pole in its mouth, and perhaps there were indeed men at the other end of the pole. But I only lately have come to realize that the men the artist saw were not repelling an attack at all. What they were really doing was hooking an alligator out of its hole. The artist shows them pushing, but I believe that his horror impaired his interpretive ability and that his "sketch from life" shows not defensive action by harassed Indians but, for the first time, the use of the familiar gator hook.

A gator hook is simply a long pole with a hook on the end, traditionally used in Florida to haul an alligator out of its den. To use it the hunter finds a gator den, probes into it with the pole, and when the alligator bites or thrashes at the pole, the hunter jerks the pole and lodges the hook somewhere about the alligator's person—preferably in his jaw. The hunter then draws out the alligator and, when his head appears at the surface, dispatches him with an ax. There is no way of determining the time of origin of the gator hook, but it is an old tool, and I am now willing to bet that the first settlers didn't think it up at all and that the Seminoles didn't either. I believe the technique was handed down from earlier times and that Le Moyne saw it in operation or perhaps got a garbled account of it from somebody else and badly missed the point when he made his drawing.

Of all the early observers of alligators, the most fluent and authoritative was William Bartram. Bartram was a sound naturalist, a warm humanist, the father of American conservation, and an able stylist in the romantic prose of his day. The heritage he left us in his second edition of *Travels Through North and South Carolina, Georgia, East and West Florida* is almost too good to be true. What happened between Bartram and alligators is something of a puzzle. The man scares the living hell out of you about them. You come away sure that either alligators have dwindled pitifully in spirit since Bartram's time or they frightened him so badly he saw visions.[3]

But that is at the first reading. If you go back as I have done and re-read all the alligator passages and ponder them, you realize that there is nothing badly amiss. A part of what has seemed to some to be unreliable reporting is simply the author's style; and the other is his misfortune, or good luck, in having twice come upon alligators in extraordinary concentrations. Bartram's day was the time of the literary style known as romantic naturalism. To write of a subject as stirring as alligators in words that failed to carry strong emotion would have been considered illiterate. If you take the trouble to go through the book carefully and assess each separate alligator encounter, and consider that it was 1780 then, long before the .22-caliber rifle was invented; that the most shattering adventures occurred in the month of May, when male alligators rut, bellow, swell, and hiss in an awe-inspiring way; that Bartram was a Philadelphian and a Quaker, quiet and peace-loving by nature and all alone with the teeming bull gators of those almost primordial times—if you think all these things together and are fair-minded person, you will scoff less confidently than some have scoffed at Bartram's alligator stories. It seems to me a pretty fundamental issue, this question of the primordial character of the alligator, important both to our appreciation of Bartram's legacy and to an understanding of the natural history of *Alligator mississippiensis*.

Considering the dramatic size and appearance of alligators, and the conspicuous place they have held in the southeastern landscape, the natural history of the species is really very inadequately known. An alligator looks a little like a big rough-backed lizard, but the resemblance is only superficial. A great many features set off the crocodilians from other reptiles. Lizards are modern reptiles, much more closely related to snakes than to alligators, whose ancestors split off of the dinosaur stem far back in the Mesozoic, some two hundred million years ago. You can see a trace of hopping dinosaurs in the small front legs of an alligator as compared with the powerful hind legs. A distinctive crocodilian feature is the spectacular voice of some of the species. Voices are not prevalent among reptiles—maybe the dinosaurs sang, but this may never be known. Another special crocodilian feature is an arrangement of respiratory plumbing that allows the creatures to hold on to large prey in the water without flooding the breathing passages. Other innovations are a diaphragm between the chest and abdominal cavities and the partitioning of the heart into four chambers instead of the three usual for the class Reptilia. In most reptiles the ventricle of the heart is a single muscular chamber, and blood coming into it from the body is mixed with newly oxygenated blood fresh from the lungs. This seems a surprisingly sloppy arrangement and may account for some of the lethargy of some reptiles—though it must not really be much of an incubus, because some reptiles have stuck with it since the Paleozoic and have done rather well, in their way. Besides, all reptiles are not lethargic by any means. In any case, the partly developed partition in the ventricle of the heart of the alligator does keep the pure and impure blood separated so that only the former is sent out to the tissues. This would appear to be a great advantage in their metabolic affairs, and it may be, though alligators spend a surprising amount of time just waiting. The blacksnake, which has only a three-chambered heart, is a spry, restless, fast-moving animal, whereas crocodilians, in spite of their clever ventricular septum, pass hours at a time drifting balefully about in ambush or lying in passive stacks on mud bars. I don't know exactly what this means, if anything.

The jaws of an alligator are so strong that gators goaded into biting steel pipes have been known to crush their own teeth or to drive them up through the top of their skull. Even the bite of an infant alligator is no light matter for the bearer of the bitten finger. In spite of the great closing power of the jaws, however, the muscles that open them are weak. You can hold shut the jaws of a big alligator with your hands, and this Achilles' heel is indispensable to the people who practice the hair-raising sport of alligator wrestling.

Another distinctive crocodilian trait is the care the female shows her nest and young. A few lizards take care of their eggs during incubation and even tend the young, in simple ways, for as long as the young ones are willing to stay where they hatch. The skinks, for example, seem particularly solicitous of the welfare of their offspring. I saw an example of this in my own yard. I lifted a board and beneath it found a female blue-tailed skink curled in a neat cavity in the soil. Surprised that she failed to streak away through the leaf mold after the habit of her kind, I poked her inquiringly with a finger, and she changed her position slightly. Then I could see that under her belly there was a little clutch of tiny skinks, three of them, all shiny and beady-eyed and with the brilliant blue tails of the juveniles of their kind. I remembered that when Dr. Robert Mount was a graduate student at the University of Florida he found parental care in another kind of skink, a subterranean species, and one of the smallest members of that worldwide family of lizards. The female of this tiny creature builds little nest chambers deep in the ground and stays there tending the eggs and grooming the little lizards for several days after they hatch. A few other lizards and at least one kind of snake curl around incubating eggs, but most reptiles recognize no parental obligations at all. So the crocodilians doing so is a mark of some distinction.

The natural growth of the alligator seems to be a little more than a foot a year. The rate is probably slightly higher down in southern Florida, where life processes are not drastically slowed by cold, than in the other southern states, where alligators go into partial hibernation in winter. On the average, both sexes appear to reach maturity when they are five years old and about six feet long. From there on, the growth rate of the males must be higher than that of the females, because they wind up with a marked disparity in size. In fact, the greater size of the male is just about the only even semireliable way to tell apart the sexes of pairs when you see a pair in prenuptial dalliance. There really don't seem to be any other external sexual features. Even Ross Allen, who practically invented alligators, has trouble telling the males from the females.[4]

Nobody has an idea how long alligators live in nature. Looking at a ponderous thirteen-footer, overgrown with algae and scarred by decades of .22 bullets and precourtship brawls, you would say he was as old as the pond you see him in. The fact is, however, nobody ever has kept account of the years of an alligator's natural life from the egg to death from old age, if indeed alligators ever die that way. So when a Cracker man assures you that some Old Joe or Old Mose that he used to know about was pushing three hundred when he got killed, the man doesn't really know what he is talking about. As far as I'm concerned it could be so. But Ross Allen has accumulated data that suggest that the maximum age may be no more than sixty years. Whether this applies to alligators free in the wild, and not just to captive ones, is not yet known.

Whatever the life span may be, an alligator is not likely to meet a violent end if it can avoid hunters. Mature alligators have almost no natural enemies. Throughout their geographic range they are the dominant predators of their habitat, and their great size, heavy armor, and frightful biting power make them immune to attack by any other animal. Even looking back into the Pleistocene, when the fauna of the southeastern states was more ponderous and aggressive than it is today, it is hard to think of any real, habitual enemy harassing the grown-up alligators of the time. If you look about East Africa for ideas on this subject you hear a few tales of fights between lions and crocodiles, or of a hippo chopping a brash croc in two, or of an elephant hanging one up to die of shame in the crotch of a tree. But on the whole, even in the parts of Africa where crocodiles are abundant and big mammals come down together at the river bank, contacts are on the whole strangely uneventful.

Baby alligators, on the other hand, are both feeble and succulent, and they are preyed on by a great variety of enemies. Otters, raccoons, herons, anhingas, snapping turtles, fish, and water snakes are all avid eaters of little alligators. Even big bullfrogs sometimes swallow them. One afternoon when I was in charge of a freshmen zoology laboratory I heard a yelp out of a girl in a corner and went over to see what her trouble was. The class was dissecting bullfrogs, and I found the girl out of her seat and pointing helplessly down at the head of a little alligator that stuck out of the incision she had just made in the stomach of her frog. It was a common bullfrog,

Rana catesbiana, the girl was working with. Little alligators have also been found in stomachs of *Rana grylio,* the smaller, more aquatic sulfur-belly frog. If it were not for the asylum of the den pool, where they live under the watchful eye of the mother, it is hard to see how little alligators could survive in numbers great enough to keep their race alive, and it is the same with the eggs in the nest. Individual female alligators vary in the attention they give their nests during incubation, but most of them spend a good deal of time there, and their presence must be an important factor in keeping egg-eating mammals away. There is good evidence that, at least in some cases, female alligators, and crocodiles too, help their young ones emerge from interment by digging the roof of the nest away when they hear the young croaking inside.

Alligators apparently have a strong sense of locale. Their movements have been studied by Robert Chabreck in the Louisiana Wildlife Refuge and Sabine National Wildlife Refuge.[5] He found that the young spend the first year or so of their lives in or near the mother's den pool but then leave home and start traveling, retaining their wanderlust until they are sexually mature. This moving about of the young alligators no doubt helps keep the population dispersed in an orderly way and prevents injurious overlapping of feeding ranges.

Once they are grown they make dens of their own, take up private freeholds, and presumably defend territory from encroachment by other alligators of either sex. The territory of the bulls is greater than that of the females, especially in the spring, when they wander in search of mates. Apparently they sometimes strike out across dry country to join a female in a distant pond when she answers amorous bellowing. It is at this time that alligators show up on Florida highways or in the yards or swimming pools of suburban homes. The Alachua County Sheriff's Office gets several dozen calls about these alligators each season and has found it necessary to appoint a special deputy to catch and rehabilitate them.

Besides their regular seasonal movements, alligators travel about to avoid drought or abnormally high water, or to congregate wherever any unusual feeding opportunity may present itself. This is, of course, a part of the reason that feeding alligators in public places, or where livestock ranges near water, is ill advised. Besides drawing unnaturally heavy numbers of alligators and stimulating competition among them, the feeding gives the gators two bad unprofitable ideas: that people are not so unpleasant after all and that anything falling into the water at the feeding station is meant for alligators to eat. Anything, one muses, or anybody.

One of the interesting results of the Louisiana tagging study was the discovery of what appears to be a strong homing urge and direction-finding ability in the alligators there. Of a lot that were marked and then moved to new localities and released, twenty-nine were later recaptured. Most of these had traveled in the general direction of home, and ten had gone all the way back to the place where they had originally been caught. One made a trip of eight miles in the homeward direction in a period of three weeks. Another was recaptured in its home lake four years after it had been moved and released twelve miles away. Still another made an almost unbelievable homing return of twenty miles. The guidance mechanisms used in such homing through supposedly unfamiliar territory are almost wholly unknown. The feats of the Louisiana alligators suggest that species may be an important experimental subject for students of animal navigation.

Anybody who spends time where big alligators range free in spring and early summer will see exciting snatches of their courtship and mating, but a connected story of what goes on has not been put together. In northern Florida, courtship activity starts at the time bellowing begins—in March or April. The complete function of the vocalizing is not known, but one thing it almost surely does is bring the sexes together. Scent glands located on each side of the jawbone and beside the cloacal opening probably also help in the pairing process.

During courting time there is a great deal of swirling and splashing about. Part of this is fighting among rivals, but part is certainly sexual play between male and female. One difficulty in observing alligator courtship is, as I said earlier, that there are no reliable external signs by which to tell the sexes apart. The sex organ of the male is internal and is everted only at the moment copulation occurs. Although males are usually bigger than females, this is not always the case. So when you watch a pair of alligators causing a disturbance it is sometimes impossible to tell who is doing what to whom.

I once watched such a pair out at Bivens Arm, and while I would bet they were courting, I would not rule out the possibility that some sort of ceremonial combat encounter between males was going on. Both the gators involved were big—one about eight feet long, the other at least eleven feet. All they did, for close to half an hour, was repeat a single curious maneuver in which the bigger one, which for narrative convenience I'll speak of as the male, would glide up beside the other, the assumed female, and then sidle up close and bend his neck sideways across her neck. As he did this he opened his mouth, and at the same time the smaller alligator opened her mouth too. Then the two of them sank slowly out of sight, and for a while bubbles and mud came up at the spot where they had gone down. For a few seconds the patches of bubbles spread, and then, one after another, the two alligators came to the surface again, twenty or thirty feet apart, drifted quietly for a few minutes, and then repeated the process in exactly the same way. They did the same thing over and over again, for a half hour at least, and then suddenly they swam separately away. Although I don't know what they were doing, their behavior had an air of instinctive ritual about it, like something all alligators must do from time to time, and, indeed, I had seen snatches of the same ceremony before.

John Hamlet, a naturalist at Weeki Wachee Springs, says that before the gators there mate, they pair off and the male (he says it later turns out to have been the male) rubs the female with his chin, and nips at her neck before her prenuptial coyness is overcome. This is probably a part of the same process I saw, but the fact is, nobody has told the whole story of alligator courtship. It is also unknown whether the spring mating serves to fertilize eggs for that same season or for later seasons. But in any case, when June comes the female builds her marvelous nest and lays her eggs in it.

Most crocodilians, like other reptiles, bury their eggs in the ground—in sand, soil, or debris. The nest of an alligator, however, is a heaped-up mound, an elaborate structure that requires a lot of instinctive skill and patience to build. The nest seems adaptively designed to raise the eggs above a changing water level and is thus a sure sign that the alligator is an age-old dweller in swamps. The American crocodile, on the contrary, digs a hole for its eggs in sand, as sea turtles do; and this suggests that the crocodile line may have had marine, estuarine, or fluvial ancestors for a long time back.

The nest of an alligator may be built in shallow water, in marsh, or on the shore of a pond, lake, or river. The first move the female makes is to prepare the site. If it is on land, a clearing is made in the underbrush; if it is in the water, slush, trash, and water plants are scratched up or bitten off and carried by mouthfuls to raise a platform above the surface. In this clearing or on this platform the alligator piles up more debris and gnashed-off or rooted-up plants until a mound three or four feet high has grown. As the stack gets higher she crawls back and fourth across the top of it, evidently to compact the material to a density that her instincts tell her is suitable for incubation. After a great deal of this bringing in of nest material and mashing of it into proper solidity and shape, the alligator scratches out a cavity in the top of the heap and lays thirty to seventy eggs in it. She covers these with mulch, crawls back and fourth across the mound for another while, and then finally returns to the water. The finished nest is a mound that may stand four feet high and have a base eight to ten feet across. On land you find this in a small clearing, and an access trail usually leads down to an open pool where the alligator spends much of her time during the incubation period.

These watch-pools are one kind of "gator hole." Gator holes are deep places in a marsh, swamp, or shallow body of water. Usually they are located where the alligator can swim away from them to open water, but sometimes you come across gator holes off by themselves, where drought has lowered the water level. In such cases the alligator has to walk away to forage. A nest built in a marsh appears as a rounded hummock surrounded by a ring of water eight to fifteen feet in diameter. Trails go out from this in various directions, but there is usually one well-marked way that leads through the weeds to the nearby watch-pool.

Apparently some alligators spend much of the two-and-a-half month incubation period near the nest, repairing any damage storms may do and finally helping the young out of the nest when they hear them croaking inside. Female alligators vary in their nesting behavior, however. Last spring I came upon the nest of a young female who was under six feet long; the nest was built on a mud flat among cattails and pickerelweed out on Lake Alice, a pond at the edge of the campus of the University of Florida. The nest was an unpretentious little pile, but the builder clearly felt some concern for it, because I could see her drifting twenty feet out from the shore and eyeing me. She seemed too little to be mature, but the nest was obviously hers. I went away without disturbing her. The next day I returned and found the nest unchanged and the gator there slowly retreating through the pickerelweed as I approached. Two days later I visited the nest again. This time the eggs had been dug out and were scattered about on the mud, some broken, some whole. I figured a raccoon had raided the nest, and I looked around for coon tracks, but there were none. The only prints on the mud were alligator tracks, and these were many and fresh. I could only conclude that the mess had been made by the alligator herself in a misguided effort to improve her nest.

In digging and moving earth and debris an alligator uses all its resources. It chews, breaks, and carries roots and trash in its jaws; shuffles and throws stuff about with its snout; pushes and piles with its webbed hind feet; scratches and scrapes with all four feet; and sloshes and sweeps with the powerful tail. The watch-pools of the female are not alligators' only kind of excavation. Both sexes dig holes to deepen the water in marshy habitat that is subject to changes in level, and both sexes may sometimes make horizontal dens. I can find nobody who has seen exactly how an alligator digs its hole in a flat bottom, but Edward Avery McIlhenny, who spent a great deal of time watching the behavior of alligators at Avery

Island, Louisiana, saw them digging their horizontal dens and described the process carefully.[6] According to him, the alligator faces the bank or the wall of its hole and starts scratching at the lower face of the bank, kicking the dislodged mud backward and sweeping it aside with its tail. This forms a cavity that slowly grows inward and undercuts the bank. According to McIlhenny the mud is never attacked with the jaws—only scratched out by the feet. When roots are encountered, however, they are ripped and yanked out with the jaws with such force that bushes and small trees over the undercut bank may sway and shake from the tugging at their roots. All this of course muddies the water badly, and it soon becomes impossible to follow the excavation procedure, except to note that every so often the alligator boils out backward from the growing den in a startling eruption of mud. The screen of muddied water that digging alligators set up is one reason so few people since McIlhenny's time have seen details of alligator excavations.

All animals to some degree affect the landscape they live in. More than most, the alligator creates landscape. It builds topography, slows or speeds up hydrologic cycles, and markedly influences the actual look of the land, as well as the organization of its biologic communities. To have lasted through the ages that alligators have been on earth has taken great resilience. A part of the resilience of the alligator is its habit of controlling its environment instead of just being pushed about by its foibles. The nests, gator holes, den pools, and trail systems of alligators all modify the environment in ways profitable to the alligators, and all affect the look and evolution of the landscape. Individual alligators sometimes live a long time in one place. Some gator holes appear to have been handed down from one generation to the next, and this process may in some cases have gone on for hundreds or even thousands of years. Some holes come to be surrounded or flanked by spoil banks, and on these, plants different from those of the surrounding region take root. Alongside ancient gator holes you often find little islands with grass, bushes, and even trees growing on them, and with herons nesting in the trees. Turtles, water snakes, otters, coons, and swamp rabbits rest on the islands when the gator is not in too close attendance. These islands have been made by the alligator and its ancestors, through their age-old dredgings of the den pool or gator hole and periodic heaping-up of nest mounds.

So an alligator in a pond is not just an eater of other animals in the pond, not merely the top predator of the southeastern wet country. He is an extraordinary member of the pond community. His excavations and dredging influence the relation between land and water. His droppings fertilize the water and contribute to its productivity. His comings and goings open channels and modify the successional processes by which ponds give way to marsh. In much of the territory alligators live in, the normal regimen is an alternation of wet and dry periods. These may be either periodic or sporadic in their timing or both. In the Everglades a rhythmic alternation of wet and dry conditions was, under natural conditions, a fundamental factor in the ecologic organization of the place. Elsewhere in Florida great expanses of so-called prairie may lie under water for months, or even years, and then suddenly become dry plain on which cattle graze. In some places these changes come very suddenly because of the geology of the area. Whenever the area of a Florida pond or lake is reduced alligators may save the fauna from complete obliteration. When a pond or marsh goes suddenly dry, hosts of fish and invertebrates suffocate in the hot mud, tadpoles and frogs shrivel, turtles bog down or go away; but each time this happens, unless the drought is too protracted, a fragment of the biota is saved in the gator holes. Like a Noah with reverse English, the alligator provides a place where some of most kinds of aquatic creatures can live on till the water returns and then repopulate the renewed habitat with their kind. The alligator doesn't do this in any altruistic spirit, understand; but he does it all the same. Actually, if you are of the turn of mind that looks for adaptive utility in any feature or capacity of an animal, you will surely come upon the thought that there is evolutionary advantage to the alligator in the saving of the rudiments of renewal. Alligators with an inborn tendency to dig deep gator holes would, no matter what the primary advantage of the digging may have been, reap a by-product of survival advantage from their ensuring to themselves a quicker restoration of their forage resources.

Once long ago, before I knew how to recognize a gator hole, I had an exciting time catching an alligator in one by mistake. I was a young graduate student then, recently arrived in Florida and out with five students in an ichthyology course I was assisting in. Our project that afternoon was to ramble about the neighborhood seining the remnants of ponds that were disappearing after a long period of drought. The first place we stopped at was a broad expanse of bare mud beside the then gloriously fecund marsh called Paynes Prairie. For several years before, what was then mostly an expanse of mud had been a shallow pond where lotus, maidencane, and pickerelweed grew in scattered stands. There was a diverse fish fauna there, and the usual run of other small animals of a north Florida pond. At least ten kinds of snakes lived in the place, and mud and musk and snapping turtles, and three kinds of hard-shelled streaked-neck cooters were there. Newts, sirens, and striped mud-eels were abundant, and a great array of water insects and crustacea peopled the pond. I had been out there collecting several times when the place was ten acres of shallow water. Now, however, disaster had killed off most of the fauna and had herded the rest together in one little pool set out in the center of a plain of mud that was all laid out in a mosaic of cracks. We walked out to the pool and stood there a while, sad to see it working so alive with suffocating fishes. Turtle backs were bobbing all over the pool, and ten times the expected number of snakes were sliding or swimming about or waiting quietly coiled at the water edge. They were mostly green water snakes, with a good sprinkling of banded water snakes among them, and even a few garter snakes and ribbon snakes were there. They had all moved in to exploit the easy fishing and frogging in the diminishing pool and now were all full and surrounded by cracked mud and with no idea what to do.

We had a seine with us, a thirty-foot minnow net. Teams of three of us took each end of this and began dragging it through the awesome bouillabaisse of gasping fishes. As the arch of the net advanced through the pool, it kicked up the flocculent ooze from the ill-defined bottom and got heavy and hard to pull. When the center of the net crossed the deep center of the pool it snagged solidly, and the haulers yelled at two boys we had assigned to wade behind the corkline, telling them to grope down into the soup and unsnag the leadline from whatever was holding it back.

"It's a log," one of them said. The other boy moved over and felt along the mass with his foot and said, disgustedly, "It's a log all right. A big one."

We all groaned because it meant aborting the haul to get the log out, but everybody pulled extra hard on the brail poles once more to see if we could move on with the load. This time the net lines pulled out in straight lines from where the hang-up was, and slowly the net began to move forward again.

"It's moving," I said. "Let's horse it on in."

The two line tenders floundered around to the ends of the net to help with the hauling, and ever so slowly the open angle of the seine advanced toward the side of the little pond. When the two teams reached dry ground, the firm footing there let them pull harder, and the net began to come in fast.

It is a time of great excitement in seining when the brail poles come ashore. The temptation is strong to drop them and start grabbing at the creatures that show up at the beached ends of the net. But we all knew about that vice, and this time we berated each other into keeping up the hauling until the ends of the net were far out on the bank and a huge, amorphous bundle of muck and wildlife had been safely dragged ashore. Then we threw down the poles and ran eagerly to reap the harvest of hapless small creatures that were flapping, slithering, or simply despondently gaping in the net. I won't try to recall the roster of species dumped so dismally together there in the mass of muck, but I can say I know of no place in the world where you could get such a diverse spectrum of palustrine life as we brought out with that haul.

We set down jars and buckets beside the heap and began throwing out the larger animals—the gars, mud fish, and suckers; the snakes and big sirens; the six or eight kinds of turtles—and running our hand through the mud to sift out the smaller fishes, newts, and striped mud-eels. We were working intensely at the sorting and sifting when one of my colleagues took hold of the log at the back of the seine bag to heave it out of the way. To his surprise the log reversed the bend of its length and opened wide the most pink-white, tooth-bordered mouth any of us ever saw on a log and emitted a hoarse, coughing hiss, and we all rolled over backward to give it any room it needed, because the log in the bag was an alligator. He was a gator bigger than any of his captors, and once free of the blinding muck and aware of what had befallen him, angry at our intrusion. After some noisy planning and fumbling at the job, we bound his jaws shut with a couple of belts and took him to town. He served there as mascot at the homecoming game and after that went to live in a campus sinkhole.

But that is pure digression. The real bearing of the incident on the subject at hand is that a few days after that seining trip, while many of the animals in the remnant of the pond were still alive, hard rains came. The little pool grew to half an acre overnight, and continued rains of subsequent days soon built it back to its old dimensions. Although the animals saved in the gator hole no doubt rattled around badly in the new space the rains provided, they quickly multiplied and set up a working community into which missing elements gradually came by other, unknown avenues, and within a year the place seemed well populated. Within five years it was again a teeming Florida pond.

Down in the Everglades, if you fly over certain sections of the vast sawgrass sea at the time of low water, you see the whole ground beneath you evenly set with little circular patches of open water, all spaced with an accountable accuracy about the plain. It is a striking sort of topography, and no one seems to be sure what molded it. The most reasonable explanation that I have heard is that the little ponds are old gator holes and that the even spacing between them represents the even spacing of contiguous hunting territories of the makers of the holes. This is not known to be fact, but it is certainly not an unreasonable assumption. In the flatwoods of northern Florida there are numerous circular ponds that seem too shallow to have originated as solution sinkholes, and some of these, too, are probably old gator holes. The total effect of alligators on the aquatic ecology and lowland topography of the southeastern coastal plain would probably be hard to overestimate.

Anyone who has traveled about Florida visiting the nesting places of water birds is bound to notice that the rookeries tend to occur in trees over open water. Some people, including me, believe that this preference reflects the freedom from predation—by coons, possums, rats, snakes, and other eaters of eggs and young birds—that the water itself and the presence of alligators under the nest would provide. There may be other advantages in the choice, but any adaptation is likely to be composite. The point is, alligators eat raccoons and raccoons surely are instinctively aware of this. And freedom from coon predation is an asset in any rookery. So the situation has the ingredients for a process of natural selection, and I believe that selection has been involved in this site-choice habit of water birds.

The other day when a neighbor was flying me around over Alachua and Marion counties on our yearly count of alligator nests, we saw a patch of white out in the middle of a marsh. The pilot banked and circled at low elevation and we could see that the patch was a clot of herons—snowies or cattle herons, we couldn't make out which—nesting in a little clump of willow trees. The trees hung over a tiny scrap of open water. We could see the head and back of an alligator in the pond, and at one edge her nest stood, a brown mound against the gray-green marsh. The patch of willows curved in a thin fringe around the nest and pool and the herons were nesting in the trees. They were cattle egrets, and the only herons nesting anywhere in the whole great marsh. Partly they were there because the only trees in the marsh were those willows that had found foothold in the spoil pile thrown up by the work of the alligator and her ancestors. But another factor in their presence was surely that the egg eaters of the marsh found it nerve-racking to rob nests over an alligator hole.

If you are a confirmed tinkerer with evolutionary ecology you can follow that thought out further, pointing out that the relation between alligator and herons would even impart a dash of lichenlike mutual advantage that would inevitably have some small effect on the evolution of both parties. The trees are there because the gator built a high bank around a permanent water hole. The herons started coming in and nesting and accidentally dropping the odd egg, young bird, or fumbled or regurgitated fish into the pool, thus unwittingly aiding the gator in her foraging. And just as inevitably and unaltruistically, the gator being down there would keep off some of the coons and snakes that would climb any trees that had no gator under them and eat the eggs and young of herons nesting in such trees.

Of the many extraordinary attributes of the alligator, the most arresting is its voice. It is a sound to go with the size and antediluvian look of the creature. It is one of the great animal voices of the earth. There are three great voices left in the southeastern United States, where the elk never bugles and the cry of the loon is hardly heard. In post-Pleistocene Florida, in the vestigial southeastern wilderness of nowadays, three magnificent voices still sound, once in a while, above the clamor of the times. They are the jovial lunacy of the barred owl, the ethereal bugling of the sandhill crane, and the roar of the alligator. I am grateful to fate that I live where all these voices can be heard. When our alligator bellows in our own front yard, I swell with pride as she swells with the air of her music.

The song of the alligator is a vast, rumbling growl that rolls up through the mist of warm dawns like something half sound and half shaking of the earth. One alligator singing alone is a moving thing to hear. Three bellowing in chorus seem to take over the world, to be doing too great a singing for any pond to hold and stay the same. They make you wonder what the choruses must have been like back in the days when big gators were spread about in primordial abundance and on April mornings all bellowed together to keep the boundaries of their territories known or to call for comfort from the opposite sex—or to achieve whatever it is that gators bellow for.

The voice—the voices, really, because they have more than one thing to say—has never been adequately described, and its function is not wholly known. One is tempted to think of it as the counterpart of the frog song, but this makes no sense. Among the frogs, only the male in most species does any singing; and for the land-dwelling kinds at least, an important effect of the song is to attract other frogs to the breeding assemblages. Within the boundaries of a frog pond, besides stimulating the females and guiding them to the amorous males, the song may also serve to help different species recognize their own kind and to help space the males peacefully apart. But both sexes of the alligator roar. Some people say they can tell the songs of the sexes apart, but I can't. The big alligator in our pond is a female, but I can't tell that by looking at her. But as I have said, our alligator built a nest three years ago and brought up several dozen young, and that proved that she is a female. And that alligator roars in a voice as virile and gargantuan as that of any other her size.

It is high time somebody kept some big alligators apart by sexes and listened to them through all seasons to record their repertoire. There are several separate sounds that they make. One is the juvenile "croak," as it is badly called—the *oo-rump* made by the young, beginning in the egg before they hatch out, and then after hatching to signal the female to come help dig them out of the heaps. Then there is the rumbling bellow of the mature alligator and a repertoire of shorter snorts and coughs. Besides these, there is a curious tenor bellow that I have heard made by an adolescent alligator in our pond. The singer is the four-year-old offspring of our big one. I judge he is the only child that has been able to keep from being eaten or driven off by his mother, though my wife believes this judgment is irresponsible, and she could be right. He is about four feet long, and in making the sound I spoke of, he seemed to be trying to bellow in a grown-up way. It was not the juvenile croak at all that I am speaking of but a loud, rolling treble facsimile of a roar. Several times we heard him singing that way one year in April and May in a patch of maidencane where he was lying. I talked with Ross Allen about this and learned that he has heard it many times, but I can find no reference in print to the singing of the prepubescent alligator. In addition to their vocalizations, alligators also sometimes make a great racket slapping the water with their tails. The only indication that this may be a form of communication is that when one alligator does it, another will often do it too.

The most completely unexpected vocalizing I have heard from alligators was the choral croaking of forty-odd young ones in a deep tree-bordered sinkhole. They, too, were the offspring of the old alligator in our lake. They all started croaking together in a chorus one afternoon, exactly like a chorus of unknown frogs. My son Tom heard them from back in the pasture where he was working and went down to investigate the unprecedented singing in the sinkhole. Then he came running to fetch me, without telling me what it was we were going to hear. When we got within range of the sound he made me stop and listen. The sound welling up out of the steep-walled sink meant absolutely nothing to me, and I figured I was aging badly not to be able to think what kind of frogs were there. We moved in closer to the pond, and I could see that the singers were spread about the surface of the pond in a patch some thirty feet across. They were all bobbing up and croaking and then going straight back under again. Each one would pop up out of the top of the water, go *roop* a couple of times, and then go under. Thirty or forty of them were doing this together, and the effect was, as I said, precisely that of a chorus of singing frogs. I had never heard a chorus of little alligators before, and have never heard one since. I have no idea why the offspring of our alligator should have all been out there croaking together that afternoon.

The voice of the alligator is the voice of the Mesozoic, a song come down through two hundred million years, from a time when there were no men, no mammals even, and no birds in the skies. The alligator was here in the Southeast to test the courage of the Spaniards when they came; and long before them the first wandering Indians drew back, appalled at the sound of his song. The alligator is just too old for us to lose. He is older than the land he lives in—he has molded the lay of that land, has shaped the look of its ponds and marshes, and even the forms and senses of the creatures he has lived with there. Saved in natural balance with a few unruined bits of the old landscape, the alligator is a relic of immeasurable value, a world treasure in our charge.

A SUBJECTIVE KEY TO THE FISHES OF ALACHUA COUNTY, FLORIDA

Drawings by Larry Ogren

Dedicated to that imposing array of naturalists, etc., who have sought to instruct their fellow men with keys. To their fellow men who have tried to use them. To all ichthyologists, anglers, aquarists, and gourmets. To anyone who sees some point in fish.

Foreword

It is to be expected that a certain amount of unfavorable comment will be directed toward the author because of his occasional departure from rigid objectivity in the present paper. Opposition to any pioneering endeavor is inevitable.

The zoological key of today is perhaps the most unstimulating and oppressive of all literary forms. Stripped of all the more succulent verbiage, its style dessicated and uninspired, and its technical arrangement often so cryptic as to render it well-nigh incomprehensible, the key may not, in any sense, be called good reading.

In the following key the author has tried to appeal not merely to a limited coterie of highly specialized scientists but to all men of all stations. It is his belief that any merit that his offering may claim lies in this quality of universality that he has sought to instill into it.

Yes, far from feeling any necessity for apologetics, the author dares predict that unless keywrights give ear to just such unconventional and, to some, treasonable innovations as the present work displays, the art of the key will wither and be borne away on the breezes of oblivion.[1]

The Fishes of Alachua County, Florida

1	Fish fossil-like, with a patently Paleozoic aspect.	2
	Fish not particularly Paleozoic in appearance; with a more Mesozoic or modern mien.	5
2(1)	Mouth with an unpleasantly toothy smile, or at least a strikingly formidable cast.	3
	Dentition not disturbing; mouth ineffectual; caviariparous. The Sturgeon—*Acipenser brevirostris*	
3(2)	Mouth with a sardonic leer and too many teeth to conceal.	4

From Archie Carr, *Dopeia*, 1941, vol. 3. Used with permission.

Mouth stubborn and determined, but not leering; teeth large and too many but skillfully concealed.
The mud fish—*Amia calva*

4(3) Leer fortunately no longer than rest of head.
The Short-Nosed Gar—*Lepisosteus platyrhincus*
Leer intolerable—twice as long as rest of head.
The Long-Nosed Gar—*Lepisosteus osseus*

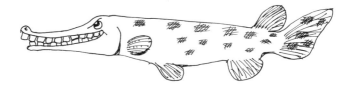

5(1) Body invariable obscenely nude; skin always awfully slimy and unpleasant to touch; eel-like (snakelike), or catfishlike.   6

Skin prudently concealed by lots of chaste scales; also slimy, but not eel-like or catfishlike; basslike, perchlike, minnowlike, etc., etc.   14

6(5) Difficult to hold; adept at tying knots in tackle; eel-like.   7

Not as above; fins with dangerous spines; catfishlike—in fact, a catfish.[a]   8

7(6) One or two pairs of feeble and utterly useless legs present. Not a fish; will not be mentioned here.

No limbs present; somewhat eel-like; an eel.
The Eel—*Anguilla bostoniensis*

8(6) or 19 End of tail lunate, bilobate, crescentic, deeply emarginate, terminally notched, or merely forked; in brief, vaguely similar in outline to a slingshot handle; if you came from 6, go to 9; if you came from 19, go to 20.

End of tail not as above, its middle part the last to arrive; if you came from 6, go to 10; if you came from 19, go to 27.

9(8) Quite pretty for a catfish; silvery with casual dark spots; brow narrow, however, and unintelligent looking.
The Southern Channel Cat—*Ictalurus lacustris punctatus*

---

[a] Any damn fool knows a catfish.

Subjective Key to the Fishes of Alachua County

	Perhaps not quite so pretty, but at least no spots; bluish, bluish-silvery, silvery-bluish, or silvery; brow very broad and unintelligent looking. The Blue Bullhead—*Ictalurus catus*
10(8)	Silvery again, but this time heavily and irregularly overcast with spurious and uncalled-for blotches of black, the ensemble with a leprous cast pleasing to some but not to me; belly fish-belly white. The Marbled Bullhead—*Ameiurus nebulosus marmoratus*
	Color normal catfish-brownish, yellowish, or blackish; belly fish-belly white or catfish-belly yellow. **11**
11(8)	Effect about what you deserve: painful and unpleasant and prolonged as a throbbing ache; if you are still interested, note that the hind end of the nubbin behind the big top fin is not stuck to the back.* **13**
	You will not wish the effects to be recalled to you; try to locate where you threw the specimen; if several days have elapsed, secure a fresh one and attempt to regard it objectively;† adipose fin confluent with caudal. **12**
12(10)	Color a stagnant brown to faded black; 3 of the nasty side spines laid end to end would be longer than the distance from snout to top fin and horrid to contemplate.

*To aid in separating the forms to follow it is almost mandatory that the reader carefully jab either of the side spines of the specimen into the fleshy part of his thumb, recording his sensations in detail. The writer disclaims all responsibility for results in the event this course is pursued.
†If you are certain the second specimen is the same species as the first, the instructions given above may be disregarded.

Schilbeodes gyrinus—Madtom, or, better, Son of a - - - - -*
Color a sickening madtom yellow, often mottled; side spines a third as long as the head and 8 times as unpleasant.
Schilbeodes leptacanthus—Madtom, or, better, Son of a - - - - -

13(11) Not spotted; more or less catfish yellowish, sometimes darker; anal rays roughly two dozen, surprisingly catfishlike.
Ameiurus platycephalus—The Flat-Headed Bullhead

More spotted; anal rays around 17; more catfishlike.
Ameiurus natalis—The Bull-Headed Flathead!

14(5) With no such structure as that described below   **16**

With the last rays of the m'n t'pf'n extended into an absurdly long filament in obvious imitation of the lordly silver king or tarpon, to whom this fish is related (tho distantly), and who is said by Breder (1929) to use this curious appendage, known in the vernacular as whip, flag, or banner, in his prodigious leaps into the air, during which, by clamping the undersurface of the filament against one side or the other, the m'n t'pf'n is diverted to port or to starboard, thus determining the side on which the tarpon, or tarpum, as he was sometimes called by the Seminoles, strikes the water. Obviously the thing is of no earthly use to the timid creature you now have in hand. Altho its stomach is gizzardlike, it is utterly unair-minded, and a whiff of undissolved oxygen always throws it into a swoon. However, if the structure is present you're on the right track.   **15**

15(14) Scales and rays of anal difficult to count; 54 and 31, respectively; expression vacuous, jejune, and oddly kippered, as of some delicatessen in-delicacy.
Gizzard Shad—*Dorosoma cepedianum*

Scales and rays of anal hell to count: 43 and 24, respectively; expression as above: similar forms are often found in olive oil, vinegar and brine.
Curator VanHyning's Banner Shad—
Signalosa petensis vanhyningi

---

*This form has been given numerous names by various workers. Most of these are either preoccupied or highly profane, and have been relegated to the synonymy of the milder Son of a - - - - -.

16(14)	Two t'pf'ns present: the main with spars, and abaft of her the mizzen, with spars but no booms.*	**Act 4**

	The same old reliable rig: a single s'l aloft, the m'n t'pf'n.	**17**
17(16)	Mouth small, low, and adapted for sucking; suckerlike. The Sucker—*Erimyzon sucetta*	
	Mouth sometimes small and sometimes large, but never suckerlike; located on the end of the fish (where it belongs) or on top of the fish (which is almost as bad as underneath), but never suckerlike.	**18**
18(17)	Canvas without spars, only batten; running altogether under balloon jibs, spinnakers, and the like.	**19**
	Fore-and-aft f'ns with spars, or at least with a small jury mast before the m'n t'pf'n.	**34**
19(18)	See couplet 8a. See couplet 8b.	
20(8)	Mouth highly conspicuous and obviously used for catching things; fore-and-aft rigging much abaft of amidship; scales too numerous to count; muskellunge-like, snooklike, barracudalike, and strikingly reminiscent of a pickerel. The Pickerels	**21**
	Mouth not as above; rigging more amidships; scales too small to count; not like any of the fishes the above is like; minnowlike.	**22**

*Spars and booms are sometimes called spines, and battens, soft rays.

21(22)   Scales not over 110.
         The Bulldog Pickerel—*Esox americanus*

         Scales about 125.
         The Chain Pickerel—*Esox niger*

22(20)   An'l f'n overrigged, with 11 or more battens.   **23**

         Rigging well balanced.   **24**

23(22)   An exquisite form, widely known among fish
         fanciers.* Beautifully colored, with a broad
         lateral band of purplish, or blackish fading
         into purplish, or bluish fading into orangeish
         somewhere and reddish elsewhere and all fading
         in formaldehyde; the dark lateral band somehow
         becomes a spot at the base of the tail; it brings
         a fair price in New York; this gemlike creature
         has been unimaginatively named.
         The Big-Finned Shiner—*Notropis hypselopterus*

         The largest, grossest, and most suckerlike of our
         minnows, and what is more, it has no tail-spot
         and half the time not even a band, but is good
         bass bait.
         The Florida Golden Shiner—*Notemigonus
         chrysoleucus bosci*

24(22)   L. l. comp.†   **25**
         L. l. incomp.‡
         The Spotted Shiner—*Notropis maculatus*

25(24)   Mh. le., is ll. ct. eg. bh. ft. of ee.§   **26**

         Mh. sl., at. no. 11. ct.∥

         The Pug-Nosed Minnow—*Opsopoedus emiliae*

26(25)   The Iron Colored Shiner—*Notropis chalybaeus* #
         The Coastal Shiner—*Notropis roseus* **

*Fanch-fish fanciers.
†Since this section of any key is tedious, the author has seen fit to abbreviate, whenever possible, for the sake of dispatch. Thus L. l. comp. means lateral line complete.
‡Lateral line incomplete.
§Mouth large, its lateral cleft extending behind front of eye.
∥Mouth small, almost no lateral cleft.
#Anal rays 8.
**Anal rays 7.

27(8)   ♂ with the an'l modified to form a more useful structure; ♂ is active, athletic, but belligerent and histrionic; ♂ is highly satyriac, his libido almost as untrammeled as the guppy's —in fact the guppy belongs in this couplet, too; liebespiel long and elaborate and fully appreciated by the ♀; ♀ more passive, philosophical, and patently pregnant, practically popping with a dozen or so fry at all times; 'v'v'v'p'r"s.   28

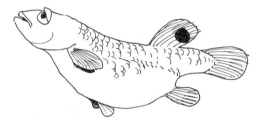

'V'p'r"s; fin not modified, but ♂ accepting his lot with fortitude; ♀ occasionally pregnant, but less radically so.   30

28(27)   A large, round black spot on m'n t'pf'n; minute but phenomenally fecund.
The Least Killifish—*Heterandria formosa*

No such spot; no such size; no less prolific.   29

29(28)   Fond of mosquitos, the females, and any kind of water. Said to rain down; probably untrue. Not as below, t' pf'n' battens fewer than 13.
The Mosquitofish—*Gambusia affinis holbrooki*

As above, but not said to rain down so often; t'pf'n longer.
The Sailfin—*Mollienisia latipinna*

30(27)   Either (condition 1) the largest of our cyprinodonts, with more than 14 dorsal rays, literally hundreds of small black dots on sides, and a singular sheeplike expression, or (condition 2) the tiniest of our cyprinodontids, with fewer

	than 14 rays, a large, rounded spot on side, base of tail, or both, and expression too small to see.	**31**
	Some other condition.	**32**
31(30)	Condition 1: The Seminole Killifish—*Fundulus seminolis*	
	Condition 2: The Ocellated Killifish—*Leptolucania ommata*	
32(30)	One broad or several narrow black lines h'll from bow to stern.*	**33**
	None: The Golden† Topminnow—*Fundulus chrysotus*	
33(32)	The former: The Red-Finned Killifish‡— *Chriopeops goodei* The latter: The Eastern Star-headed Minnow— *Fundulus nottii lineolatus.*	
34(18)	Vent and ventrals both abnormal, the former under the chin§ and the latter with rays I, 7. The Pirate Perch—*Aphredoderus sayanus*	
	Vent ventlike, normally located—a typically percoid vent; ventrals under no circumstances I, 7: I, 3, perhaps, or I, 5—even I, 4 1/2,‖ but never I, 7: all straight percoid, even sunfishlike, unless you have somehow got hold of one of those silly little hangers-on that has no anal spines and doesn't belong here at all.#	**35**
35(34)	A.=D.; i.e., A., VII ±I, 5–18.	**36**
	A.D.	**37**
36(35)	Spines of D. VII to VIII; often makes a fine panfish. The Speckled Perch—*Pomoxis sparoides*	
	Spines of D. XI to XIII; finicky; almost never makes a fine panfish. The Flier or Paneluder—*Centrarchus macropterus*	

*Waterlines.
†Often off the standard.
‡Also its mate.
§Many of our foremost authorities agree that this form can never tell whether it has hemorrhoids or a sore throat.
‖Half is sometimes lost somehow.
#The Flagfish—*Jordanella floridae*.

37(35) See couplet 8. Disregard the traffic signals to the right; come right back here; if section a. impressed you as the most pertinent, go to **40**

If you rather lean to b., go to **38**

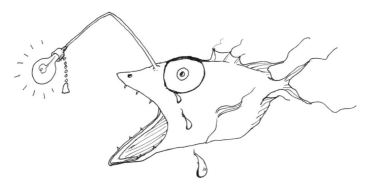

38(37) Top of head with an anteriorly directed filament that extends far ultra-cephalad and terminates with a luminous expansion; eyes immense but weak and watery; mouth much larger than the fish. No county records; you may have got your data mixed; your keying is also sloppy.

Everything under control. **39**

39(38) Spines of D. IX; small, but will bite worms; use a tiny, long-shanked bait-hook and snap the line sharply at the slightest nibble; sides sometimes with glorious blotches of royal purple.
The Blue-Spotted Sunfish—*Enneacanthus gloriosus*

Small; not available to the angler; a striking aquarium fish if you put mud in the bottom; on a sand bottom he turns pale like sand, but with mud he often shows glorious blotches of royal purple; spines of D.V. to VI; a veritable pygmy among sunfish.
The Pygmy Sunfish—*Elassoma evergladei*

40(37) Ein tiefer vorliegendedornigevomhintern-weichenteilabteilender Ausschnitt in die Ruckefinne.* **41**

Notch in D., if present, not deep. **A**

---

*German. In view of the elusive subtlety of the German composite and the ever-threatening danger of loss of value thru translation, the writer has deemed it the safer course to leave the statement of this important character in the Mother Tongue.

41(40) Maxillaire plus longue, sa partie postérieure derrière de l'oeil.*
The Largemouthed Black Bass—*Huro salmoides* (Lacepede)

Boca más pequeña maxillar casi no llegando al margen posterior del ojo.†
The Smallmouthed Black Bass—*Micropterus dolomieu*

A(40) With a constellation, or, better, nebula of small black pigment spots on posterior part of D.; if these are close enough together to resemble a spot, okay; if not, you'll have to go Jordan and Evermann.‡
The Bluegill Bream—*Helioperca macrochira*

AA No constellation or better nebula as above; maxillary micropterus-like, ending under pupil; range, eastern United States, from the Great Lakes to Florida and Texas, west to Kansas and the Dakotas; common in South Carolina.
The Warmouth Perch—*Chaenobryttus gulosus*

AAA No blue on head, and where the ear should be. The spot's not lengthened horizontally . . .   **AAAAB**

AAAA The head is oft bedecked with lines of blue; The spot is longer here, and narrow too,

Oh, noticeably longer here, unless Your fish has not begun to adolesce.

If so it be that juvenile you've caught Then you have struggled through the key for naught;

For every ichthyologist agrees That young sunfish are all the same species.§   **AAAAC**

AAAB (AAA) P. pointed and extending beyond base of last anal spine; a reddish margin on posterior part of opercular spot in life; an exceedingly succulent form: the most eminently edible of our fishes; secure as large a series as possible

---

*French. In honor of Lacepede, who described the form.
†Spanish. For comparison.
‡Jordan (D.S.) and Evermann (B.W.). *The Fishes of North and Middle America*. A descriptive Catalogue, etc., with 392 pl., 4 vols.; thick octavo. The most important systematic work on our fishes; in great demand and difficult to secure; binding lacking or somewhat skotched. Wash., 1896–1900. $40.00.
§It is opinion of the writer that the ordinary free-verse couplet, used almost universally among keywrights, is a distinctly inferior art form. The above paired rhymed lines of iambic pentameter are the heroic couplet of the classicists. This form is introduced for comparison with the dull, sloppy, and effete versification so characteristic of zoological keys.

and filet them; lay the salted and peppered
filets meat-side down in 5 mm. of excruciatingly
hot olive oil in a heavy iron skillet; no cornmeal, please; cook one minute on each side;
sprinkle with paprika and lemon juice, and go
to town! A prime panfish; a peerless platefish; a
perfect palatefish.
The Shellcracker—*Eupomotis microlophus*

P. shorter and more rounded; no red on ear
margin; side scales each with a small black
spot often; only average edibility; an ordinary
cornmeal and lard fish.
The Speckled Bream—*Sclerotis punctatus
punctatus*

AAAAC(AAA) Check scales in 4 or 5 rows; ear-spot
light-margined; will cooperate with the fly
fisherman; the correct lure is a Maudlin Mamie
or Mae West body with vermillion shackles,
pink panties, and crimp-cut brassiere, tied with
mauve silk on an H. P. Spick hook with selected
tapewormgut snell; dust lightly with lilac bath
powder and use only between 4 and 5 A.M. on
Washington's Birthday.
The Red-Breasted Perch—*Kenotis megalotis
marginatus*

Check scales in 7 to 9 rows; ear spot dark to
the edge; recommend tackle schedule; No, 741-D.
or 741-L: our superb Bully Boy unslacked line, or
our one-armed Finnegan Ketchemall dynamite.
The Stumpknocker—*Lepomis auritus solis*

Act 4(16)

Enter the two-topfinners: the Brook Silversides, *Labidesthes sicculus vanhyningi*; the Mullet, *Mugil cephalus*; the Crawl-a bottom,
Hadropterus nigrofasciatus; the Swamp Darter, *Hololepis barratti*;
and the Brown Darter, *Villora edwini*. Their expressions are smug
and their double dorsals proudly spread; they are obviously pleased
with themselves and sing in chorus a strange little song that is
somehow vaguely pathetic.

All: We are the fishes with two top fins!
Notice how each of us winks and grins!
We're not much conceited and don't like
   to boast,

> But we're fully convinced that we're better
>   than most
> Of that lowbrow bunch whose D.'s are single;
> So we keep to ourselves and don't deign to
>   mingle
> With pot-bellied killifish, gars, and jack,
> And the rest of the proletarian pack.
> Hotchacha and a hydihy!
> We're the boys and girls of Tau Tau Phi!

(*They join pectorals and start to dance.* Hadropterus, *who is slightly intoxicated, trips over* Labidesthes' *anal.*)

Hadropterus: You clumsy ass. Why the hell can't you keep your—Say! (*he looks at the fin sharply*). What's your anal formula?

Labidesthes: (*timidly*) I, 23.

The Darters: I, 23!

The Mullet: What a nerve! A mug like him trying to get chummy with people that's got class. Why, you skinny little sardine, I'll punch your preopercle.

(Labidesthes *exits hurriedly,* L. C., the Mullet, *in pursuit.*)

Hadropterus: Good riddance. I always thought there was something fishy about those guys.

Villora: While you were talking I looked at that vulgar Mullet's anal, and would you believe it! His anal formula is III, 8.

Hadropterus and Hololepis: III, 8!

Villora: Yes. And (*in a shocked tone*) they say he sells for three cents a pound and goes good with grits!

Hadropterus: My God!

Hololepis: Sickening. I am sorry you've had this experience, Villora. But they have gone now, so don't bother about it any more, dear. And now Hadropterus, may I have a few moments alone with Villora? There's something I . . .

Hadropterus: Sure. Good luck old man. (*He exits.*)

Villora: I like him, don't you? He's different somehow. I suppose it's his extra anal soft rays. He has 9 or 10, you know.

Hololepis:	And we both have 7, don't we, dear?
Villora:	Of course.
Hololepis:	We have a great deal in common, sweet.
Villora:	I suppose so. But there's that funny little spot at the base of your caudal, and you have lots more scales than I.
Hololepis:	Oh, I don't know. How many have you?
Villora:	About 39, the last time I counted them.
Hololepis:	I thought so. I have only a dozen or so more. That's nothing, my dearest. Don't you think you could learn to care for me, Villora?
Villora:	Oh, yes, Hololepis. Sometimes I'm afraid I already care for you too much. But . . .
Hololepis:	But what, honey?
Villora:	Well, you see . . . You see, Hololepis, I'm afraid your family would never accept me. Yours is one of the oldest names in the *Percidae,* and we—well, you know we weren't described until 1930.
Hololepis:	Why you silly little girl. As if that could make any difference. My family would never think of impugning the validity of one of Hubb's species, no matter how recent.
Villora:	(*warming*) Oh, you are sweet. But darling, we're different genera.

Hololepis: Don't be Victorian, dear. That's no objection at all. Inter-generic crosses—and happy, successful ones—are common. Look at the sunfishes. (*He takes her into his pectorals and looks into her eyes searchingly.*) You do love me, don't you, dearest? Please say you'll hybridize with me. Tell me, Villora, will you?

Villora: (*returns his look tenderly, and after a moment, says softly*) Yes, Hololepis. (*They embrace.*) But what on earth will the children be like?

Hololepis: They'll be hell to key out, honey.

Hound Magic

T. J. and I were loafing on the fire line at the Down Tree Stand and the dogs were strung out trailing in the pines south of Green Bear Hole when old Trouble jumped his deer. In ten seconds the other Walkers and the Julys were on his neck, and the puppies and long-eared dogs were piling down fast in anticipation of the hot scent ahead. They bunched and the chorus took shape, and the wild music bearing down on us set T. J. and me to trembling.

My friend T. J. is one of those lucky Central Florida grove men who took up fruit growing mostly because it wouldn't cut too deep into their hunting time. He's been through twenty seasons in the Ocala Scrub—more counting the old August hunts—but he still trembles when the dogs come down. The still hunter gets over his ague after a deer or two, but the dog man never does.

T. J. and I crouched, both of us trying to back into the scant cover of the same little holly bush. "They're pushing him out here," he whispered. "The bucks run that ridge when they strike in Green Bear Hole and the wind's northwest. He's going to run right over us." He kept fingering the action of his .351, and his hands were shaking, but I knew they'd steady down when the time came. I hoped mine would.

There was never any doubt as to how that deer was running. Both of us knew his course and where it would bring him out as surely as if he'd been following a turpentine road. The only doubt was when he would burst out of the bushes into the little patch of low rosemary where our split-second chance to shoot would come. From the pitch of the dog's excitement we judged they were pushing him, maybe tailing him no more than a hundred yards, maybe four hundred feet.

As the pack seemed practically on top of us and we knelt with gun butts at our shoulders, sweating, we heard the deer. He was right where he

From *Field and Stream* 53, no. 2 (1949).

ought to be. He cracked a stick and ripped through a bush and slapped a palmetto fan and our fingers took up trigger slack. Then there was silence—unbelievable, almost unbearable quiet where the brush should be exploding; where the sleek red side of a December buck should be blotting out the view down the cooter-furrow, there was nothing.

The muzzle of T. J.'s rifle sank and he began whispering curses under his breath. "What happened?" I asked idiotically.

"Be quiet," said T. J. "Let's see what them dogs does."

"There's nobody north of us," I said despondently. "He'll almost surely head for the breeding ground if he's a buck."

"He's a buck all right," said T. J. "Let's wait and see how he runs."

The dogs came on. The clamor that had been a strong whine a few seconds earlier was now the pointless yodeling of a gang of deluded mutts, for the buck was out of the block. We sat in the misery of anticlimax and waited till the mob reached the spot where we had heard the deer. Trouble was still in front, though the rest were pushing him hard. As he hit the board-stiff leaves of the palmetto through which the deer had jumped, he stopped giving tongue, and one after another the dogs in the following pack did the same. Up to the palmetto bush they were running a deer on a head-high scent that they drank in like hot soup; beyond the bush there was no trail, and their coursing chant died out with the failing deer smell.

The Walkers started tearing around in a frenzy, some overrunning hopelessly, their conviction tapering off in a maze of older tracks, some crying and yelping as if they were being whipped. The puppies gave up; the ones that winded us came over to visit, and the rest flopped to pant on the cool sand where the trail died or trotted off to hunt skunks. The long-eared hounds cruised slowly about, meticulously searching out the foot scent, snuffling their noses against the sand and pine needles, every so often throwing back their heads and bellowing out a heartbroken protest at the injustice of it all.

"Well," said T. J., "I'll be a suck-egg mule."

"T. J.," I said cautiously, "did you ever see anything like that before?"

T. J. made as if to throw his .351 down on the sand to stomp on it, but he got a grip on himself in time. "Did I ever see anything like it," he yelled. "Why you know dern well I've seen the same dern thing, and right here on the Down Tree Stand. I can tell you when I first seen it and how many times I seen it, and it'll be nearly half the times I've took this stand! Every time it happened I tried to forget it, because it don't make no sense. There's no sense in it and no justice, and if anybody ever catches me on this stand again I hope they kick me right square in the ass."

I saw he was just going to get worked up if he kept talking, so I interrupted. "Well, me," I said, "I don't half believe what happened, and I'm going out there to see what kind of hole the deer fell in."

T. J. sat on the down tree and began filling his pipe. I started walking the sixty feet out through the waist-high rosemary, past a scattering of pines and a scrub oak tangle. With no trouble at all I located the palmetto bush with incoming tracks on one side and, just beyond it, hard to make out under the skidded dog tracks, the scrabbled sand where the deer had fallen all over himself stopping. After some backing and filling I also picked up the tracks the deer had made after the check, where he had wheeled and struck out at maybe forty degrees with his back track. I trailed out a piece of this leg, and the only dog tracks on it were those of the cold trail hounds walking the ground scent.

"What a hell of a note," I thought. "Here comes a deer drilling north straight and fast through the skinny pines, pressed some by the dogs but with nothing much to worry about. Suddenly he hauls back on his hams, scared stiff, apparently by his own shadow. He plows up three yards of sand stopping and then strikes out again toward the southwest. Worse than that, and this was the part that just didn't fit, when the dogs overran the angle they could never pick up the scent again. The deer was still in the drive after all, but he might just as well have been in the next county!"

I was beat. I started walking back toward the stand, puzzled and mad. When I reached the palmetto I looked up ahead, and there through a latticework of little-leafed oak limbs, I saw T. J. sitting on the old down tree, morosely smoking his pipe. Get that. I saw T. J. plain as day, and his rifle against the log and the puff of blue smoke he was letting out. "T. J.," I yelled, "can you see me?"

"Hell, no," he said. "Let's go ketch dogs."

Right there I began to wonder if there isn't more to this scent-running business than meets the eye; if maybe a hunting hound doesn't have more to contend with than just following a more or less potent stink.

When I got to talking around I found a couple of old timers who had noticed the same cockeyed and improbable thing on the Down Tree Stand that we saw. But apparently nobody ever got around to walking out to where the dogs checked and looking back toward the stand. When I did, I learned why deer often turned there. They could see the stander! The down tree wasn't any proper stand at all. Humiliating as it was (and still is) to admit, a curious screening effect of the intervening foliage made the deer invisible as he approached the point where he would break into the open, but left the stander wide open! The deer only had to look in the stander's direction, which in our case he did.

But what in the world was the connection, if any, between the emotional shock the deer received on viewing our ugly faces and the petering out of the trail?

Plenty, apparently. I've done plenty of hunting since that day in the scrub, and I could tell a dozen tales with the same plot. I've talked to a lot of hardheaded and unimaginative hound men who recognize the symptoms, and though they don't usually like to discuss them, they sometimes will.

More than that. I have had the pleasure of reading Joseph Thomas' great book *Hounds and Hunting,* and in his unique chapter "Eccentricities of Scent" I found just what I was looking for. Besides assembling more observations on the vagaries of varmint trail that anybody ever got together before, Thomas describes the temporary failure of scent in time of crisis very clearly. He even quotes from Masefield's poem "Reynard the Fox," in which a boy's terrier waylays a passing fox, scares him, and thereby ruins the chase for the following hounds and riders. This poem tickles me so, I have to repeat a piece of it:

> And young Jack Cole peered over the wall
> And loosed a pup with a 'Z 'bite en, Saul,'
> The terrier pup attacked with a will,
> So the fox swerved right and away down the hill.
> Through the withered oak's wind-crouching tops
> He saw men's scarlet above the copse,
> He heard men's oaths, yet he felt hounds slacken
> In the frondless stalks of the brittle bracken.
> He felt that the unseen link which bound
> His spine to the nose of the leading hound
> Was snapped, that the hounds no longer knew
> Which way to follow nor what to do;
> That the threat of the hound's teeth left his neck,
> They had ceased to run, they had come to check,
> They were quatering wide on the Wan Hill's bent.
> The terrier's chase had killed the scent.

Whether or not Masefield took this particular incident from life, at least it shows that the pink-coat people also are well acquainted with the phenomenon.

After all, why shouldn't nature pinch hit in this way for one of her creatures under stress. We accept things just as odd without turning a hair. We don't give it a thought when sudden strong emotion—fear or anger—boosts our own muscular efficiency to double normal or more, in a flash draining blood from the danger zones of the skin, dumping into it an emergency ration of high-test fuel sugar, and in the same wink of an eye, raising the blood pressure to hurry the stuff to the muscles and making us breathe faster to provide extra oxygen to burn it with! All this, and a lot more, is the work of the little pair of glands, the adrenals, and it seems reasonable that a set of emergency measures as complicated as this might easily include a temporary suppression of the scent we normally exude. Maybe our own odor is less strong when we are frightened. Anyway, it is widely claimed that a dog can detect a scared human by a difference in his odor. Who knows but that the difference is a suppression of the normal odor.

This is all just conjecture, of course, and maybe I'm sticking my neck out, but I'm willing to bet that some day some open-minded physiologist who is also the kind of a dog fanatic who watches and cogitates on every move his running hound makes as if it were his first-born learning to walk, some guy such as that is going to have himself a lot of fun and maybe turn up some unexpected monkey business as well.

And actually this emergency scent control probably isn't very different from the well-known, or at least much-discussed, reduction of the odor given off by nesting birds and by animals when they "freeze."

We once had a mallard duck that nested regularly without interruption in a woods in which possums, coons, and foxes roamed unhindered and made off with roosting poultry almost weekly.

Moreover, I have watched three hounds faultlessly trail a marsh rabbit through a half mile of flooded swamp, hauling in the thread of the trail like line into a reel, right up to where the poor short-legged rabbit crouched in a form, frozen in the assurance of doom. Then I stood and grieved as a hound with twenty-four-inch ear-spread and two talented puppies made idiots of themselves in their frenzied attempt to locate by nonexistent scent a quarry that they could easily have seen if they had been built that way.

This scent failure is hard on a dog's morale, but I suppose that, day in and day out, the toughest problem the working dog faces is having to decide which way to run on a trail. Although a young dog is just about as likely to go one way as another, with a little experience he learns that there is a right way to go and a wrong way, and before long he can choose the course that pays off almost without hesitation. So much is this black magic—this right way sixth sense—taken for granted that if old Bell or old Blue should happen to backtrack for a stretch it's apt to be laid to a falling barometer or to the Republican Party or to the dog's having slept with its nose in the ashes.

This olfactory direction-finding is to my mind one of the wonders of the world, and even though it has been the subject of a lot of high-powered theorizing and debate, it still looks to me like an open question. Some will tell you it's just a matter of smelling both sides of a bush and then taking off on the side opposite the strongest odor, and this is probably good so far as it goes. A deer does hit one side of a bush harder than the other and probably does leave more scent there, and dogs certainly put in a lot of time smelling both sides of a bush. But suppose there are no bushes. What coon hunter has not seen his dogs trail true on bare mud, or even swim the trail, poking his muzzle at each small straw or floating chip to glean the shreds of sodden scent he needs to keep his roaring? No, the bush-smelling theory is at best only part of the story.

In the Florida scrub I have several times led my black-and-tan bitch to trails in the bare sand, laid by deer I had previously seen run by. I have watched her recognize by sight the formless, caved-in footprints, push her rubbery nose into one and then another, inhaling deeply through the powdery sand, and after snuffling back and forth and maybe back again for a distance of several yards give tongue with matchless three-tone cold-trail cry and strike out dragging her ears along the deer trail and almost without fail in the deer's direction.

What was it that drove old Rhoda the right way? What was she testing when she snorted up and down the twenty-yard stretch? What is a trail, anyway, and when you answer that tell me how a piece of it can have a back and a front!

The obvious explanation, or rather the only suggestion that seems to have any merit at all, is that the off end of a trail segment, being younger, is stronger than the on end. In other words, the dog works back and forth till he decides which way the scent grows and takes that way. Well, maybe so.

But before swallowing this explanation whole, let's look for a minute at the statistics of the thing. Let's say that when the stretch of trail in question was made, maybe five hours previously (and that's a pushover for all but the hottest-nosed coursers) the deer was moving at a conservative lope of, say, fifteen miles an hour, or twenty-two feet per second. The test section used by the dog for orientation is perhaps a hundred feet long (and I've seen them decide in half that distance), and at the rate the deer was going, this strip of scent required a little more than four and half seconds—make it five—to lay down. In other words, if the near end of the segment of trail is five hours, or eighteen thousand seconds, old, then the far end is all of eighteen thousand and five seconds old. And old Drum's job is not just locating this infinitesimal ribbon of odor; he's got to find the new end of it, when "new" means five seconds' difference in tracks made five hours ago by a buck who is now pulling acorns in the high green back of the old camp in the Ero drive!

If this is so, and I don't see any way around it, then the only decent thing to do is to cultivate some new respect for old Drum because he rates it.

Of the three kinds of trail scent—that left in footprints, that brushed off on bushes and such, and that which vaporizes directly into the air from the body of the moving animal—the air trail is the one the hounds love, the one they run most freely, heads up, eyes half closed, mouth open; in the joy and savor of the scent not even bothering to mouth and round the syllables of their cry, just beating it out with their breath, *aa, aa, aa, aa.*

The air scent can be tricky, though. Every old hound man has seen a trail, streaming out behind a running fox or deer, picked up by drifting air and set down intact a dozen or a hundred yards to leeward, and the dogs down there running it like lunatics, happily ignoring the fact that they are covering ground the quarry never touched.

I saw the most soul-stirring piece of the offset trailing of my experience when Neal Adams' dogs jumped a big tom bobcat in a gum hammock in North Florida. The twenty-eight cat-trained Walkers were a deadly machine, but no wildcat lacks guts and cunning, and this old tom was a heller. Things seemed to be going well enough, though, when suddenly Neal, who had moseyed down into the hammock a way, came charging out and hollering blood-curdling abuse of some degenerate who had strung a wove-wire fence with two top strands of barbed wire smack through the middle of the cleanest hammock in Levy County.

Now a wildcat knows what wire will do to trail-crazy dogs as well as you do. Neal Adams knew too. He started kicking a rotten pine log apart, busting out the lightard knots and handing one each to my wife and me and his three other guests. Then he told us to run down the fence line leaving a man behind every so often while he scrambled around in the hammock trying to head the dogs and get a shot at the cat with the little .22 rifle he was carrying.

The other three guests were on the stout side, and Marge and I outstripped them, gaining thereby the doubtful privilege of unhooking four shrieking freshman puppies hung up by their belly-skin on the murderous top wire. Before we had freed the last pup, the cat turned back into the fence three hundred yards further on, leaving more dogs hooked and howling in agony and rage. And before we even reached those the devilish varmint had hung up another batch.

Seven times within a matter of forty minutes that cat crossed the fence, raking the pack with the barbed wire, and this despite our frenzied shuttling back and forth, brandishing of our pine knots, and yelling. He was too slick for us. And anybody who tries to tell me he went through the fence all those times by pure chance has got three strikes on him at the start.

It seemed for a while as if this combination of wire fence and old tom bobcat might not only wreck the hunt but butcher the best pack of Walkers in Florida. But little by little the veteran hounds—the grim old dogs with plenty of wire scars but no very recent ones, who take wire like Percy Beard on the high hurdles—closed the gap behind the racing wildcat. The seventh time he took them to the fence the cat's margin of safety was too small for comfort, and he must have begun to get just a little uneasy, because after he had crossed he bored straight out at right angles to the fence and into the gentle steady wind that was pushing through the clear understory of the sweet-gum forest. Far out into the woods he raced at smoking, scared-cat speed, and then when he began the swing that would cut back and put the dogs through the meat chopper again the old cat undid himself. As he streaked along, angling back toward the wire, I suddenly caught sight of him for the first time. Like a scared gray hant I saw him sweep past the tree trunks two hundred yards or more upwind. I saw him and cussed Neal Adams for making his people leave their guns at home, and then I forgot to cuss because the dogs came swarming down.

Down the gum-corridors they rioted, devil dogs broke loose from hell, the smell of their own blood strong in their noses but stronger still the steaming cat-scent. As they boiled down among the trees their song came before them, reinforced by their pace, coarse mouth and turkey mouth all alike now, no single note distinguishable in the pounding blood-chant, a solid mass of sound rolling throughout the woods from a solid mass of cat-death.

Then I saw them, too, and all of a sudden felt sorry for the cat, for I could see how his end was coming, and I knew his race would run out in a matter of seconds. He was running a wide arc—upwind, crosswind, downwind—just now still on the crosswind leg almost parallel to the fence but starting to cut back. The dogs, too, were running an arc, almost exactly concentric with the wildcat's course. But while the cat had passed me at maybe a hundred and fifty yards out in the woods, the dogs were going by no more than sixty yards away!

They were taking the easy air-trail, the cat-meat trail, and they didn't care a hang that the wind had shifted it a hundred yards or so. Neither they nor the poor old tom knew that the inner arc was bound to cut the outer before the fence was reached, but it did. And I heard the yelp of the first fast, simpleminded puppy who skidded up and made the first grab as the bobcat tried to come to bay! I ran out with my pine knot, but when I got there the cat was gone, buried deep under a seething half-ton of insane Walkers.

The old tom bobcat was as smart as they come, and they come shifty, but he never knew what a lousy trick the crosswind played on him, and I'm glad. When he washed out it came quick, and right down almost to the last I think he was savoring the thought of the next bunch of yearling cat-killers he was just about to hang up on the top fence wire.

Carnivorous Plants

Back in the days when science fiction was unsophisticated, a standard way to make people uneasy was to introduce the subject of carnivorous plants. The idea of a vegetable preying on animals has an eerie flavor, like the conceptions of cartoonist Charles Addams. When the plant has spiked jaws that grab hold of attractive human blonds, a reader or viewer is—or was, in the old, naïve days—reliably stirred. The thought of being eaten by a lion or a crocodile evokes fright. Being ambushed and swallowed by a snake or embraced by an octopus raises tension still higher. But to imagine oneself becoming the repast of a plant generates a level of revulsion not easy to explain.

There was a time in my life when I believed there were vegetables that ate people. The thought frightened me. It gave me a sense of insecurity comparable to that evoked by seeing John Barrymore change from Jekyll to Hyde when I was six years old. My elders took pains to assure me that neither of these things ever really took place. But Mother didn't seem too sure about the plants, and the memory of a picture I had seen somewhere—of a maiden being engulfed by a big-jawed tree—stayed with me and left scars on my character.

Today the thought of plants eating people is a little less disturbing. Science fiction and special movie effects have exposed us to such shattering concepts that a maiden being eaten by anything is taken pretty much in stride by even the prepubescent public. Nevertheless, there remains a mystique about animal-eating plants that fascinates people; and for botanists they have been of great scientific interest for as long as their habits have been known.

Some people used to doubt that the animals trapped by plants were actually digested and used as a source of nutrients to support metabolism and growth. This doubt was dispelled by Charles Darwin, who was fond

From *Animal Kingdom* 86, no. 6 (1985).

of insectivorous plants and wrote a book about them. To prove they were true carnivores, he fed small cubes of roast lamb to a sundew. The sundews are a varied lot of very beautiful little plants, the leaves of which are set with spikelike hairs. Each hair is tipped with a glistening droplet that is irresistible to an insect but, like flypaper, mires it on contact. Darwin's sundew digested and absorbed the roast lamb eagerly. To test its digestive skill still further, he placed tiny drops of milk among the spikes, and these too were digested. Darwin continued his experiments by feeding sundews all manner of substances derived from animals and found the plants able to cope with them all.

That carnivory is a successful enterprise for the plants that practice it is proved by its occurrence in widely different kinds of plants. It is accomplished by mechanisms ranging from passive capture—in which the victim is mired in sticky secretions or enticed into pitfalls from which exit is impossible—to active trapping, as in *Dionaea*, the Venus's-flytrap. The flytrap entices victims into the reach of the two halves of its spike-bordered leaves, and these snap together when sensitive hairs are stimulated.

In one case insectivorous members of the same family—that of bladderworts and butterworts—have developed different ways of trapping prey. Butterworts use the flypaper technique, but the bladderworts, which are mostly aquatic, swallow little animals into a tiny sac with a trapdoor in one side. When sensitive hairs are stimulated the trapdoor slaps open, and low pressure that is somehow maintained inside sucks in the prey, like the slurp gun that collectors use to catch little reef fishes.

The prey of carnivorous plants is not restricted to insects. Various other invertebrates are taken, and frogs, mice, and little birds have been recorded as victims. Nevertheless, I detect a curious tendency among authorities on carnivorous plants to refrain from using the word prey for the animals they capture. But what more must the plants do to qualify as predators? Roar? Slaver? Gnash teeth? It seems to me that since the plants prey on bugs, any way you look at it the bugs are prey.

Once, driving up to Tallahassee from the Gulf coast of Florida, we came upon a stand of pitcher plants—genus *Sarracenia*—for which that region is famous. The Florida Panhandle and adjacent parts of Alabama and Georgia have more species of *Sarracenia* than any other area, and their mixed stands in the acid flatwoods there are among the great shows of decorative herbs of the United States. They usually bloom in April, and the flowers are fetching; but it is the leaves—the particolored hooded pitchers, some of them up to three feet tall—spreading through the open flatwoods like imaginary plants, that steal the show.

We stopped the car near the spectacular congregation to see how many kinds were there, but we couldn't decide. Three species certainly, but there could have been twice that number. Pitcher plants hybridize, I am told, and turn up in different phases. We soon gave up trying to sort them out and looked around for other acid-land plants. We found sundews and two kinds of butterworts all over the place. There were at least two types of terrestrial bladderworts, too, and these started me wondering whether and how the bladderwort's trap works when the plant is growing on flatwoods soil instead of in water.

As we drove on to Tallahassee we passed several more tracts of pitcher plant woodland, and at the edges of some we noticed patches of tall-stemmed white flowers we didn't recognize. When we got to town we learned that these were Venus's-flytraps, which are not supposed to occur anywhere near Florida. They surely were planted there by somebody who knows good habitat for vegetable carnivores.

Carnivorous plants were far more widespread in Florida before the flatwoods were drained, slash pines planted in rows, and wildfire controlled. Whenever you came to a patch of acid ground you were likely to find butterworts, sundews, and pitcher plants; I used to wonder what the connection was between acid soil and carnivory. I eventually learned that it evolved simply as a way of avoiding competition.

Acid soil is poor; the acidity hastens the leaching out of nutrients—most seriously, salts of nitrogen and phosphorus. Any plant that can survive in such a habitat shakes off a lot of crowding. Pitcher plants, sundews, bladderworts, butterworts, and the Venus's-flytrap have all found a way to live in it: by getting nutrients from insects. In bogs and patches of burned-over sour ground they are free of the danger of being crowded out by the plants that require soil richer in nitrogen and phosphorus.

The insectivorous adaptations of pitcher plants have had a lot of attention from botanists. The pitchers are the traps. Their vivid coloration is presumably an adaptation to attract insects. They secrete a nectar that draws the insects into the mouth of the pitcher. Downward-pointing hairs on the upper wall of the pitcher ease the descent of the victims onto a slippery area, and there they lose their footing and fall into a little pool of liquid at the bottom of the chamber. The pool contains enzymes that digest the prey, and it has been proved that the products of digestion are absorbed directly by the cells of the leaf.

Apart from the insects that get caught are creatures that have evolved the ability to go into the pitchers and scavenge the corpses of the prey. It seems a perilous way to make a living, but various species of bacteria, protozoans, rotifers, nematode worms, copepod crustaceans, mites, and insects all are able to survive in the liquid in the pitchers. So a pitcher plant also houses a small ecological community.

Durland Fish and Donald Hall set out to explain how three kinds of dipterous insects avoided competing with each other as habitual occupants of such a tiny ecological enterprise.[1] By some ingenious experiments, Fish and Hall showed that the flies avoided stepping on one another's toes, as it were, by coming into the pitcher community at different stages of its development and by feeding in different ways—one on new captives floating on the surface, another on debris suspended in the pitcher pool, and a third on corpses at the bottom of the chamber.

When Fish was a graduate student at the University of Florida he impressed the botanical world by adding a member of a different family to the list of carnivorous plants and by showing that it too has taken up carnivory as a way of escaping competition. The new one was a bromeliad, a member of the pineapple family.

Most of the epiphytes—the plants that perch on tree limbs and trunks —of the American tropics are bromeliads. Epiphytic bromeliads hold pools of water in the axils of their leaves and get nutrients from rainwater that has leached down through the forest canopy or from leaf litter that falls into the axillary tanks. In the tropics a great many kinds of vertebrate and invertebrate animals have adapted to life in bromeliad tanks, and some are committed to spending their lives in them. The droppings of these guests help make up for the absence of soil.

Fish studied the bromeliad fauna of Florida as his doctoral research, and in his dissertation he recorded the discovery of an insectivorous bromeliad.[2] The new bug-catching kind, which inhabits the uppermost levels of the treetops, where no other bromeliad grows, is *Catopsis berteroniana*. Its leaves are erect and tubular, and they hold rainwater at their bases. Insects that visit these little tanks are unable to escape because the walls above them are covered with fine chalklike powder that, as in pitcher plants, gives no foothold. These powdered surfaces also reflect ultraviolet light, and Fish believes that this property is an insect attractant. When he took four of the bromeliads home and fastened them to fence posts in his yard they prospered, catching 136 insects belonging to eight orders in eight days.

While looking over what I had written about man-eating plants having been a theme for horror stories of yesteryear, I began to wonder how the old tales got circulated so extensively. Rummaging in the library, I learned that the mother lode of the stories was a letter written in 1878 by Carle Liche, a German traveler, to Omelius Fredlowski, a Pole. This account was subsequently widely published—and widely believed—in Europe, America, and Australia.

It took an interlibrary loan for me to get a copy of the letter (in *Madagascar: Land of the Man-Eating Tree*, by Chase Salmon Osborn) and when I did, all my old anxiety came back. The following passage from it clearly reveals why the tales of man-eating plants spread abroad and scared people and lived on into the time of my youth. The excerpt describes the behavior of a meat-eating tree during the ceremonial sacrifice of a woman by natives of Madagascar:

> The atrocious cannibal tree that had been so inert and dead came to sudden savage life. The slender delicate palpi, with the fury of starved serpents, quivered a moment over her head, then as if instinct with demoniac intelligence fastened upon her in sudden coils round and round her neck and arms, then while her awful screams . . . rose wildly to be instantly strangled down again into a gurgling moan, the tendrils one after another, like green serpents, with brutal energy and infernal rapidity, rose, retracted themselves, and wrapped her about in fold after fold, ever tightening with cruel swiftness and the savage tenacity of anacondas fastening upon their prey. It was the barbarity of the *Laocoon* without its beauty—this strange horrible murder. And now the great leaves slowly rose and stiffly, like the arms of a derrick, erected themselves in the air, approached one another and closed about the dead and hampered victim with the silent force of a hydraulic press and the ruthless purpose of a thumb screw. A moment more, and while I could see the bases of these great levers pressing more tightly towards each other, from their interstices there trickled down the stalk of the tree great streams of the viscid honeylike fluid mingled horribly with the blood and oozing viscera of the victim. . . . May I never see such a sight again.

I'm sure we all share that hope. The description continues, but I have deleted the more graphic passages. I am glad that Liche lived out his days in the nineteenth century. It is awful to think what today's special-effects technologists would have done with the material he could have given them.

The Cold-Blooded Fraternity

Our farm, with the woods around it and the marshy pond out in front, offers a fine sampling of the main groups of reptiles. The farm is no cretaceous swamp, but through the years we have caught, seen, stepped on, or had the house invaded by a great many different kinds of reptiles, representing all the main groups except the tuatara.

Most people have a vague feeling that no reptiles except turtles are to be trusted, and so they make no effort to find out anything about them or even to learn what they are.

A proper reptile, to begin with, is a vertebrate animal. It has scales, breathes air (not water), characteristically lays shelled eggs, and depends on outside sources for its body heat. There are in the world only five main groups of animals that fit this definition. They are the turtles, the lizards, the snakes, the crocodilians, and the strange, little-seen creature called the tuatara, which looks like a lizard but is not.

Our farm is extraordinarily blessed with turtles, and a stroll around it gives a good, broad impression of turtle architecture and customs. The turtles there are both aquatic and terrestrial. They spread in size from the four-inch stinkjims to snappers that weigh thirty pounds or more and in disposition from the long-necked, short-tempered, carnivorous soft-shells that lurk on the pond bottom snorkeling air through tubular nostrils to the various placid, dome-shelled water turtles, called cooters in Florida, which bask on logs, forage for edible trash on the bottom or cruise about chomping at fallen cow-crazy petals floating on the water. Industrious gopher turtles follow their small trails out from vaulted burrows and graze among the grazing cows. Once in a long while a box turtle comes meditating through the yard and retires to its shell until we go away.

The specialty of turtles is armor, a shell made up of a top part (the

From *Life Nature Library: The Reptiles,* by Archie Carr and the Editors of Time-Life Books. Copyright © 1963 Time-Life Books, Inc.

carapace) and a belly part (the plastron). The two parts are centrally joined at each side by a bony bridge. Top and bottom, the shell has two layers, an outer one made up of broad, horny scales joined by stout seams and an inner, usually much thicker layer of tightly jointed bones. Because the seams of these two mosaics do not coincide, the whole structure forms an extremely strong casing within which much, and in some cases all, of the turtle can be safely stowed away.

Turtles range in weight from a few ounces to well over half a ton. The biggest turtles are aquatic, but there are big ones on land too. The famous Galapagos tortoises, and others on islands in the Indian Ocean, have reached weights of more than four hundred pounds.

There is a popular notion that turtles live, as it were, forever. Little is known scientifically about their maximum life span, however, and it is not possible to evaluate this notion. Careful sifting of records from zoos, and of the generally shaky evidence afforded by turtles with dates carved into their shells, has led some herpetologists, as students of reptiles are called, to suggest a figure of a hundred years as a probable maximum. Few turtles living near people realize this potentiality. They get run over on the road, their marshes or ponds are drained, their streams are poisoned, or they are simply caught and eaten. Fortunately for their survival, their longevity does not mean they are slow to mature. In fact, turtles reach sexual maturity in a surprisingly short time. In the several species for which data are available breeding may begin at ages of from three to eight years.

From the standpoint of abundance and diversity, the lizards and snakes are by far the most flourishing reptiles of today. Between them the two groups include about six hundred genera and at least fifty-seven hundred species. They occur on every continent except Antarctica. Snakes are clearly derived from some ancient kind of lizard, and the two are put together in the order *Squamata*.

One of the features distinguishing the lizards and snakes from other reptiles is a drastic reduction of bones in the temporal region of the skull, which reaches its extreme among the snakes. Another is that the anal opening in lizards and snakes is transverse, instead of longitudinal as in crocodilians and turtles. Finally, both snakes and lizards have paired copulatory organs, and both have distinctive sets of sensory cells, called Jacobson's organs, in their mouths.

As to differences between snakes and lizards themselves, most lizards can close their eyes, but a snake's eyes remain permanently open behind a clear covering called the spectacle. The unblinking stare of snakes may account for some of the superstitious fears people have about them. Snakes also generally have a single row of widened scales under the belly, whereas the scales of lizards tend to be more nearly the same size above and below. Lizards typically have some sort of external ear; snakes have none. In most lizards the tail can be readily shed, and in some the broken-off section snaps and jumps about in an irresponsible way, perhaps to distract attackers. It is easy to imagine that this allows the rest of the lizard to slip quietly away from the scene while its attacker is preoccupied with the twitching tail. Later, a new tail generally is grown, shorter and sometimes lighter in color, with a different scale pattern from the one left behind.

The most obvious difference between typical lizards and snakes, however, is the leglessness of the latter. Although there are lizards that have no legs and that superficially resemble snakes, it is still generally easy to draw the line between the two groups. At the same time, it is helpful to keep in mind that snakes are really a specialized and quite successful sort of lizard.

Of the two groups of the *Squamata,* the lizards are of course the older. They have the conventional body plan of a typical land vertebrate: four legs, five toes to a foot, and the sprawling gait of the earliest reptiles. Most of the adaptations that have allowed them to spread and prosper are relatively unspectacular changes in the old four-legged look—exceptions being the various groups in which the legs have been lost completely. As vertebrates, lizards are a fairly representative group, and it has been suggested that the lizard would be more suitable as a type with which to introduce freshman biology students to vertebrate anatomy than the universally used frog. Perhaps it sounds cynical to say so, but I think the answer there is that the frog, being tailless, fits dissection pans more gracefully.

In spite of their fundamentally conventional body plan, modern lizards are a diverse lot. They range in length from two inches to ten feet. They may look like dragons and they may look like worms, and they show a complex adaptive range through terrestrial, arboreal, subterraneous, and aquatic environments.

Out at my farm lizards are all over the place on warm days. The large family of the *Iguanidae* is there, represented by the slender anole that stalks insects on the screens and by the scaly-backed fence lizards that bask on almost every log or stump. This is, as the name suggests, the group to which the big tropical arboreal and marine iguanas belong, and it includes a host of smaller forms. Its counterpart in the Old World is the family *Agamidae,* which has a curiously similar structural and ecological spread.

The classic lizards—classic because they are of the Old World and since early times have beguiled European naturalists—are the personable lacertas of the family *Lacertidae.* My farm is, of course, devoid of lacertas, and this is a shame. Their place is partly taken by the athletic, silky skinned, six-lined racerunner, *Cnemidophorus,* a member of the related family *Teiidae.*

The teiids are an alert and active lot of lizards that mostly occur in South America but are also found foraging about dooryards among chickens and babies in villages throughout the Caribbean. At the farm the racerunner streaks across the paths, and throughout the summer the hot, bare sand in all the open places is slashed with its tracks.

The most cosmopolitan lizard family is that of the skinks, the *Scincida*—shiny-scaled lizards with protrusible tongues and a generally surreptitious air, which may account for their being occasionally known as scorpions. One of the farm skinks, the ground lizard, is a brown slip not much bigger than an old wooden match. The other is the big, burnished redheaded scorpion, which irks squirrel hunters with the noise it makes skittering up the trunks of trees.

A day of plowing or harrowing is likely to turn up two kinds of lizards that have no legs. One is the so-called glass snake, which dashes headlong into the soil and makes one wonder why a European relative is known as the slowworm. The other is the Florida worm lizard, a blind pink double-ended creature resembling a blunt night crawler. The worm-lizard family is called *Amphisbaenidae*, which literally means "to walk on both ends," because among most of the creatures in it both ends look surprisingly alike.

The giants among modern lizards are the monitors (family *Varanidae*), which are believed to be close to the ancestral line from which snakes came. Monitors are fierce predators that run down and rend quite large prey. They have long, flexible necks, protrusible tongues like snakes, and an intense way of looking at people that makes them feel uneasy.

There are no monitors on the farm, nor indeed are there any wild ones anywhere in the Western Hemisphere. There are no real chameleons out there either, nor any Gila monsters. But the greatest deprivation we suffer is the lack of geckos. To residents of the tropics all over the world, the geckos—lizards of the family *Gekkonidae*—are an institution. Geckos have many features that endear them to the open-minded. They live trustingly in people's houses. They have loose skins, vertical pupils, no eyelids, and a voice—the only voice among lizards, and one of the few really reliable voices in the whole class of reptiles. The name gecko is an effort at onomatopoeia, being suggestive of the sounds made by an Old World species known technically (and a bit redundantly) as *Gekko gekko*.

We come now to the snakes, and it is interesting, in light of their close relation to and derivation from lizards, that some of the most primitive living snakes are burrowers whose skeletons retain traces of the old limb girdles from ancient lizard times.

Much better known, more conspicuous, and far more numerous are the four principal snake families, the *Boidae*, the *Colubridae*, the *Viperidae*, and the *Elapidae*. The first of these, and the most primitive, includes the New World boas and anacondas, and the pythons of the Old World tropics. They are constrictors: they kill their prey by wrapping themselves around it and squeezing. Although there are some small members of the group, the *Boidae* as a family are notorious for being the largest snakes in the world. There are well-authenticated records of anacondas more than thirty-seven feet long, pythons of thirty-three feet, boa constrictors of eighteen feet. Generally speaking, pythons and boas resemble each other closely, but the former lay eggs and the latter bear live young.

The second major snake family, the *Colubridae,* is by far the most diversified. It contains about two-thirds of the world's snakes, which, for want of a better name, go under the general heading of "typical" snakes. There are colubrids wherever snakes are found. As might be expected, they are the most numerous and most varied snakes on my Florida farm. Within a hundred yards of the house you can find snakes to show the main colubrid habits and habitats. For instance, if you go to see why a frog is screaming down at the pond edge you will probably find that a water snake—a banded or green water snake—or perhaps a common garter snake has hold of it. Or if it is a tiny, thin scream, it may be that a ribbon snake is trying to swallow a gangling tree frog. The most ubiquitous snakes on the place are the black snakes and the rat snakes, or chicken snakes. The chicken snake is a slow-moving, mainly arboreal constrictor, most readily found by joining any conclave of irate blue jays. The black snake belongs to the genus *Coluber,* which, with the whip snakes, includes some of the most agile and enterprising snakes in the world.

The farm colubrids range in size from a scarce red-bellied *Storeria* that creeps about in the leaf mold and matures at a length of five inches to the handsome gun-metal indigo snakes that can swallow a rabbit. The rear-fanged contingent—the typical snakes that have independently hit upon poison as a device to get food—is represented by the crown snake, *Tantilla,* which has grooved fangs but is too small to get a grip on a human. You can dig up a scarlet snake a foot underground, or chase a coachwhip over many acres, or see a slender twig on a bush turn into the air-thin grace of *Opheodrys,* the rough green snake. Most of the typical snakes on the place lay eggs, but some bear their young alive. Some eat any living prey they can catch and swallow, but the hog-nosed snakes lean heavily on toads for food; the king snake eats other snakes; and the red-bellied, shiny horn snake in the pond eats mostly salamanders.

The remaining snakes of the farm and of the world constitute two groups among which the production and injection of venom have become highly refined adaptations. These are the vipers and the cobras, the latter group including the coral snakes and sea snakes. The cobras and their relatives (family *Elapidae*) are found around the world in tropical regions. They kill prey with venom injected through fixed hollow or grooved fangs located toward the front of the upper jaw. Some of them, like the mambas of Africa, are big, swift, obstreperous, and even warlike. Others are timid burrowers or foragers in leaf mold, like most of the American coral snakes.

The poisonous snakes with the most elaborate venom-injection apparatus are the vipers (family *Viperidae*). They are found on all the continents except Australia; in fact, most poisonous snakes of temperate regions are vipers. Most vipers are stout-bodied snakes with the wedge-shaped or heart-shaped head generally thought of as the mark of a poisonous snake. The pit vipers include such imposing animals as the rattlesnakes and the tropical fer-de-lance and bushmaster. Their name is derived from a sensory depression, or pit, in the side of the snout between the eye and the nostril. This is elaborately supplied with nerves and blood vessels and is an organ specialized for detecting the presence and range of warm objects. Most pit vipers eat warm-blooded prey, and the pit is no doubt used primarily in feeding, but like the rattle of the rattlesnake, it is perhaps also of value as a means of avoiding injury under the hooves of big mammals.

The only way my farm stands out in the serpent-rich North Florida landscape is in the prevalence of poisonous snakes there. A coral snake on the lawn is no great event at all. Cottonmouth moccasins come up from the pond, and every now and then the cook kills one at the kitchen door. Diamondback rattlesnakes are abundant, and pygmy rattlers turn up occasionally underfoot. In fact, a snake found in the yard is almost as likely to be venomous as harmless, and this is by no means the case in most places.

We had five children ranging the premises for a dozen years, and they remained unbitten. The only friction we have had with snakes has been their killing of our dogs. A series of low-slung dogs has lived on the place—trustful bassets and testy dachshunds opposed to all forms of life not canine or human. Of these, three, and perhaps four, were killed by snakes.

Like any proper Florida pond, ours has an alligator in it. Alligators belong to the order *Crocodilia,* the third major group of reptiles.

Crocodilia are in some respects the most advanced of reptiles. They share with higher vertebrates a four-chambered heart, and they have a partition between the cavities of the chest and abdomen that suggests the diaphragm of mammals. In crocodilians the cloacal opening is a longitudinal slit, as in turtles, instead of being transverse, as in lizards and snakes.

Full-grown crocodilians range in size from between three and four feet, in the case of the Congo dwarf crocodile and the dwarf caiman, to top lengths of about twenty-three feet in the Orinoco crocodiles. The biggest American alligator ever measured was slightly over nineteen feet long.

Crocodilians have one of the few well-developed voices among reptiles. It is a matter of family pride that an alligator which appeared at a tender age in our pond finally began to sing. He had cruised silently about for nearly a decade, methodically eating or driving off alligators that came in from other places, and cracking mud turtles with a ghastly noise that disgusted my daughter. Then one misty morning he came forth with the earth-shaking, soul-stirring song of his kind. I suppose this means he is growing up. Or she—the females bellow too.

The remaining order of living reptiles, the *Rhynchocephalia,* has one living representative. This is the tuatara. There are no tuataras on our farm. They live only on a few islands off the coast of New Zealand.

Living with an Alligator

Until recently I have been very intolerant of people's intolerance of alligators on their property. It seemed to me that one hope for keeping free alligators in the land was their willingness to live in urban and suburban lakes and ponds. Anyway, it has always appeared to me that anybody lucky enough to have a personal or neighborhood wild alligator ought to be so gratified that he would gladly put up with inconveniences in order to perpetuate his happy state. I still hold that belief, to some extent. But now our own alligator—the oldest of the three alligators that we seem now to have in our own pond—has grown big, and I can understand the problem a little better. I am attached to our alligator, deeply so, but she has made me see that a three-hundred-pound predator in the front yard is simply bound to be a mixed blessing at best. In fact, the plight of people vis-à-vis alligators is sort of summarized in my own family's experiences with the alligator in our pond.

Living with little alligators is no problem. It is when they get to be five or six feet long and then grow on up to ten, twelve, or even more intemperate lengths that trouble may come in the yard pond or urban lake. The alligator in our pond is going on eleven years old. We have known him since he was small. He really is a female, but it is a family custom to call her "him." Anyway, this alligator is pretty close to us, really, except during the coldest weather, when he sinks morosely into the muck and is close to no one for days at a time. Shortly after he bellowed for the first time another big alligator turned up from somewhere, and for a while the two of them circled and dashed about the pond, making waves and white water that hid whatever they were doing out there together. A little later we discovered that our alligator was a female. We found a nest that she had made on the far shore of the pond, and not long afterward forty-two little gators came out of the nest. She herded them into the deep sinkhole in the corner of the pond, and for a while they all lived there together.

Our alligator kept growing in size and appetite and began eating things she had not been able to eat before. If our eight-acre pond had stayed that size it might have supported the alligator, but during dry spells it shrank, and then she would gorge herself with distressing abandon. You could hear her cracking turtles clear up in our living room. Then, after a while the pond would flood again and dilute the reduced remnants of turtles, bullfrogs, rice rats, and swamp rabbits. Hard times came for the alligator, and she stayed hungry all the time. The sulfur-bellied frogs disappeared from the open pools, and the big bullfrogs that bellow under the buttonbushes around the edge grew few and quiet before each season ended. We had six mallards on the lake, and the alligator ate them all. We got four more mallards and built a strong cage for them, half in the water and half out. On the second night it was there the alligator ripped a hole in the wire and ate two ducks and then tore another hole to get out.

After a while we seemed to have reached the point at which we must choose between having a big alligator and having a pond with other animals in it. The alligator seemed to be consuming everything, and one day we shamefacedly reached the decision that she had to go. Not killed, of course—we still felt affection for her, and anyway killing even your own alligator is against the law. She would have to be caught and hauled away somewhere to an ampler, wilder place. My wife called Sheriff Joe Crevasse and explained the problem, and he seemed not to be surprised to hear about it. It was May, the month when all through the county alligators were walking, and Joe had already had thirty calls about them being on the road or in people's yards, and he was expecting more in June. He said his staff was busy, but that they'd try to move our alligator if it was really important. My wife said it was.

The next night Deputy Shelly Downs came out to the house. Shelly was pound master for alligators and rattlesnakes in the sheriff's department. He came in a truck and brought his regular gator-catching equipment: a canoe, a long pole, a heavy nylon line, and an assortment of nooses made of different gauges of stainless steel cable, each with a shackle at one end for the noose to run in and a ring to fasten the line to at the other end. To catch an alligator, you open one of these loops out and fasten it to the line, then suspend it from the end of the pole. The idea is to get close enough to the gator to slip the noose over his head, then tighten the springy loop by a sudden heave, jerk the pole loose, and play the alligator on the line.

A big soft-spoken deputy named Jenkins came with Shelly to paddle the canoe. The two of them got out of the truck and we walked down to the shore to get the lay of the pond. Shelly played his spotlight out over the water, and at once picked up the alligator's eye shining on the far side of the pond, like a hot blown ember about to blaze. He snapped off the light and the two of them went back to the truck, put the tackle in the canoe, took it down to the shore and slid it into the water. Deputy Jenkins got in the stern and Shelly told me to sit on the bottom amidships. He pushed the canoe out into deep water and took his place in the bow, crouching low there and holding the long pole with the loop hanging from the end. Then he turned on his light and quickly picked up the blazing eyes again.

Jenkins was a good hand with a paddle. He drove the thin canoe through the floating mat of duckweed with only a whisper of a sound. The eyes of the alligator burned on unblinking in the spot of light, and it seemed no time at all till only a pole's length lay between them and Shelly's noose. We slipped over a school of bream, and they flipped and churned away in every direction, two of them jumping into the boat and clattering noisily on the bottom. Jenkins and I groped about after them, uneasy over the noise they were making, and finally found them and slid them back into the pond as quietly as we could. In spite of even this disturbance, the gator stayed there steady in the tight beam of Shelly's light, unalarmed, floating high, with most of her back and the keel of her tail above the surface of the water. It seemed sure we were about to make fast to her. She looked very big, maybe eight feet long, and I wondered belatedly how exciting it would turn out to be to play her from a canoe. I recalled how an alligator that Ross Allen and his wife Celeste had noosed had grabbed the bow of their canoe in its jaws, crumpling the aluminum hull and breaking Celeste's ankle inside it.

But Shelly seemed to have no misgivings. As we got closer he gently dipped the noose in the water and the glide of the canoe slid it toward the alligator, toward the long smile of her face, now clearly visible down there under the clear tan of ten inches of pond water. It seemed sure that our alligator was about to have the first bad time of the ten years of her unmolested life.

But just as the circle of cable slid up past the two humps at the end of the broad snout, it touched a strand of coontail moss, shifted sideways, and brushed the side of the smiling face. Instantly the pond erupted, the gator was gone, and Shelly was sitting there swearing.

"You nearly had her, Shelly," Jenkins said.

That stirred Shelly to more rough language, but he soon got quiet again and started shining his light around to see where the alligator had made off to. Fifty yards over at the edge of the pond we picked up the blaze of the eyes again. Again Jenkins began shoving us across the solid-looking deck of duckweed. Quiet as a blown leaf, the canoe slid up to where the gator lay like a log facing directly toward us and unblinking in the light. Once more Shelly pushed the pole out and let the noose come down, and it made only the tiniest of ripples as it slipped cleanly over the snout, past the bulging jowls, and back to the neck of the unsuspecting alligator. When it got past the shoulders, Shelly heaved hard on the pole and the noose tightened solidly in place just behind the short front legs. The cold steel squeezed, and it scared the alligator badly, and she dived for the bottom. Her dive slanted in under the canoe so fast that Shelly was not able to keep the pole up, and it slapped down on the gunnel and snapped and splintered there. Shelly dropped it and grabbed for the line but he missed the first grab, and the moment of slack let the steel noose loosen. The alligator streaked out free of the loop, roiling a trail across to the farthest side of the pond.

After that we stayed out in the lake for quite a while, swearing and trying to pick up the alligator's eyes, but we never found them again.

"She's on the bottom thinking about that cold steel that squeezed her," Jenkins said. "She'll be down there a week."

After a while the green tree frogs in the buttonbush trees on the floating islands again took up the chorus that our noise had interrupted, slowly building back up its curious cadence, one by one joining in, all around the pond, until a thousand of them again were shouting together—*"fried-bacon, fried-bacon"*—in the strange unison they work into. Shelly played his light around once more, and still there were only bullfrog eyes to see; so we went in and had some cake and coffee my wife had made to celebrate or to console us with. After that Shelly and Jenkins stowed their gear, got into the truck, and started the engine.

"We can try it again in three days," Shelly said. "No use any sooner. She'll only be wanting to hide."

Three nights later they came back again. I was away, but the two of them went out on the pond and made two good approaches to the alligator. Both times the noose snagged in the coontail just before it slipped into place, and after the second miss the alligator refused to hold for the light. Shelly said the gator was getting prejudiced against them and that a trap would be the only way to get the gator now. They went away to let her calm down before we built the trap. Early the next morning the alligator broke into the pen and ate three more mallards.

So, after all the soul-searching we did, after painfully concluding that we had to banish our alligator, we still have her out there in the pond. The water has stayed at high level for three years, and her depredations have been less obtrusive than before. I don't know what she eats, really. Snakebirds fish out there with her, and Florida gallinules each year tend small bands of utterly helpless and no doubt succulent chicks. Each fall flocks of teal, widgeons, and ring-necks come down, and on the warm days when the gator is out they swim all around her, opening in only the most inadequate-looking circle of emptiness about where she lies. The ducks seem foolhardy, but they obviously know what they are doing. In fact, they and the gallinules and herons all know so much of alligator psychology that they use her as a foraging aid. If they are nearby when the alligator starts moving through the floating vegetation, they often swim in close to her—just as a cattle egret walks up close to a grazing ungulate—and move discreetly alongside, snatching at the small creatures stirred up by the advance of the gator's bulk.

I ought not fail to mention that one of our flock of tame mallards, a big shiny drake named Sam, escaped the fate of his fellows and lived on in the pond for a couple of years, joyfully joining the flocks of widgeons in the winters and perhaps from them learning how a duck can live with an alligator. One spring Sam disappeared. Maybe he flew away with the wild ducks, or possibly the alligator ate him. But in any case he showed clearly that a fat, slow-moving, tame mallard can adjust to coexistence with a ravening predator, and this is a good thing to know. Knowing it has brought new tolerance to our attitude toward the family alligator and has steeled us to stand her misdemeanors and try to learn from her a necessary protocol for future vital relations between her species and our own.

The Moss Forest

As I write, the moss is sick. The Spanish moss—the gray, graceful decoration of the Southeastern coastal plain, boon companion of the live oak trees—has grown brown and light-bodied and is shredding away on the wind. In 1968 moribund moss began turning up in scattered places, and by late 1970 the trouble had spread in a broad diagonal belt through the upper half of the Florida peninsula, from Sarasota to coastal Georgia. Six months later it had reached Biloxi, Mississippi, and was marring Middleton Gardens in Charleston, South Carolina. It had jumped over Savannah's Bonaventure Cemetery, the finest moss garden of all, but it seemed only a matter of time till that too would be invaded.

Some people see the passing of the moss without regret—some even with satisfaction. These individuals seem to me to be out of their mind. The southeastern United States could stand some changes, but the loss of the moss is not one of these. A few days ago a man I know, a chap from the North, asked me why I cared if the moss was sick. "It's only a damn pest," he said. I fancy myself a fairly reconstructed Southerner, but what I thought was: "Yankee, go home." A man who thinks of Spanish moss that way made a bad mistake coming this far south.

I admit that moss is a mixed blessing. Like anybody else, I resent the way it prunes trees that need no pruning; and before it got sick and stopped growing too heavy to hold itself up in wind, I used to hate to have to pick it up after every storm. But Spanish moss is the most influential plant in the southeastern landscape, the most widespread, and by all odds the most distinctive. A big old live oak without its moss looks like a bishop in his underwear. Lord only knows how a whole live oak hammock would look if its moss should disappear.

Spanish moss hanging in trees has a subtle, changing look that painters of landscape find hard to capture. On clear days when the moss is dry there

Adapted from *Audubon* 73, no. 5 (1973). Copyright © 1973 by the National Audubon Society, 700 Broadway, New York, N.Y. 10003.

is something about the gradient of light and density in the pendant clumps that eludes both the brush and the pencil. Not being able, myself, to draw anything, I can't tell what the trouble is; but the attempts at realistic delineation that I see are all too stiff or fuzzy or dense or in some other ways distressing. But the freer gropings for the spirit of moss make it seem like a thing the artist has not known long and lovingly enough. The trouble is both texture and illumination. The festoons fall in soft tapers that are opaque gray above and thin out into translucent spikes below. And much of the time the moss is in motion. The light tips make thin wind gauges that dance out downwind in moving air. To a wise eye these tell the speed and direction of the wind more accurately than dancing leaves or swaying branches do. I have a friend who likes to fish off Cedar Key, sixty miles away on the Gulf coast. It doesn't take much of a wind to ruin a fishing trip out there, so no matter what the weather report has been, before leaving home my friend looks out of his window and studies the movements of the moss. If only the feathery tips are dancing he gets ready for the drive to the coast. But if the swing starts far up in the tops of the clumps he knows that the wind on the Gulf may run stronger than the predicted fifteen knots, and he thinks of other things to do that day.

On rainy days the whole look of the moss landscape changes. It is these days that make dyspeptic newcomers complain that the moss gives the land a funereal look. They grow most dejected on windless days, when long rain has wet and weighted the festoons and they hang plumb and sodden like wet hair and seem twice as long and thin as before. Because the soaking wets the tiny gray scales that hide the green of every strand, the moss turns a curious olive color; meanwhile, the trunks of the trees have turned black from the soaking and the wet squirrels have grown dark and thin under their plastered-down hair. These are the times when moss most seriously grieves the preternaturally melancholy person.

Then, when the summer squalls come or the northwesters blow, the moss gives the land still another look. It thrashes and swings with the gusts of the storm, streaming out flat when the wind holds strong, or streaking away in banners. A woods without moss stands still in a tempest and only tosses its branches about, but when moss is out there flying with the wind the woods seem to be in motion, like landscapes seen through the window of a running train. And when the storm is gone still a different look comes over the forest, because much of the moss that was up in the trees now lies scattered about on the ground.

The detractors of Spanish moss also find it hard to give up the thought that the plant is a parasite up there sucking the lifeblood out of the trees it hangs on. That is not true. After its first few weeks of life a moss plant has no roots to suck with. It probably gets a little nourishment out of the water that trickles down the bark of the tree it is growing on, but moss has its own chlorophyll and carries on photosynthesis like most seed plants. It probably gets most of its nutrients from the dust of the air, trapped by the film of water held by the mosaic of pointed scales that covers the moss plant. Spanish moss is not a moss at all. It is actually a bromeliad, a member of the mostly tropical family to which many other less weird-looking air plants and the pineapple belong. Because Spanish moss hangs down instead of standing in upright rosettes, it looks nothing like most of its kin, though a few near relatives show how the curious body form evolved from the more usual bromeliaceous condition. The technical name of Spanish moss is *Tillandsia usneoides,* in recognition of its resemblance to usnea, another gray pendulous epiphyte known commonly as old-man's beard. Actually old-man's beard is no moss either, but a lichen.

Spanish moss ranges throughout the coastal plain from Virginia to Texas; and closely related species are found as far south as Argentina. In the tropical part of its range, moss is nowhere as prevalent as it is in the southeastern United States, occurring in isolated enclaves that are spread about in a discernible pattern. Even in the coastal Southeast moss does not by any means flourish everywhere. The gaps in its distribution are not always easy to explain. It clearly thrives where humidity is high, and it dies out in deep shade. In one place or another it grows on almost any kind of tree. It seems partial to live oak and cypress, and one might say that this is because both grow best near water. Cypress is mainly a swamp tree, and live oak grows most exuberantly around the shores of lakes and bays. On the other hand, trees growing on high and dry sand ridges are sometimes loaded with moss. In Louisiana the hackberry seems almost as good a host to moss as the live oak.

When you look casually at a clump of moss it is not easy to make out the anatomy of the individual plants. Each is made up of a single exceedingly thin stem, sometimes twelve feet long or even more, with tiny pointed leaves. The whole plant is covered with a frosting of minute scales, and these hide the green of the chlorophyll and give the moss its characteristic gray color. A clump or curtain of moss is a cluster of these slender plants, some of them young and short, some old and long, all falling together in festoons. In early summer moss grows minute yellowish-green flowers, which smell faintly sweet on quiet nights. The flowers produce little seeds with feathery floats that fly them all about the country when the wind blows. When hurricanes come, whole moss plants, too, are blown far and wide.

From primitive times until only lately Spanish moss has been a useful resource. Both Jacques Le Moyne, who came to Florida with Laudonnière in 1564 to paint the curiosities of the land from life, and Jonathan Dickinson, who escaped "the cruel devouring jaws of the inhumane of Florida" after a shipwreck on the East Coast, saw Indian women using moss for skirts. It was dried and twisted into cordage, too, and up until a few years ago it found a steady market as stuffing in horse collars, and in upholstery for furniture, carriages, and, later on, cars. For decades moss picking was a lagniappe industry in Louisiana, and at one time fifty factories ginned moss in the state, and there were at least ten moss factories in Florida. During the depression many a poor Floridian made his whole living by picking moss, trapping fish, and roving the flatwoods for deer's-tongue, an aromatic herb used to perfume tobacco. In those days moss trucks were a common sight on the roads. They were usually ramshackle and always monstrously overloaded, and when they got caught in the rain their loads soaked up water and they broke down. Live moss, freshly pulled down from the trees or picked up after storms, could be sold for a cent or two a pound. There was plenty of it in the woods, and harvesting it was not much trouble. What you did was fasten a foot or two of barbed wire to the end of a long pole, shove this up into a hanging clump, twist the pole to engage the moss, and then haul it down. In prime moss woods a good picker could gather up to five hundred pounds a day, and a whole family might bring in as much as a ton.

When a strand of live moss is killed and cured, the outer sheath falls away. What remains is a hard black fiber known in the trade as hair. It took as long as six months to produce the best grade of moss hair. They used to kill the plants by piling them in mounds or packing them into pits and then soaking the moss with water. Some people preferred lye water. The moss was turned periodically to cause even decomposition of the "bark," as the pickers called the soft outside covering. When it was good and dead they hung it out on a fence or a clothesline to dry, turned and rehung it a few times, then took it down and sent it to the factory to be ginned and baled. The moss hair was frizzly, black, and springy, and in seat cushions it would keep its resilience longer than most other kinds of stuffings did. Well-cured, glossy black, and free of trash, it brought as much as six cents a pound.

Although the curing process looked pretty simple, there were hidden subtleties in it somewhere, because I was never able to turn out a batch of six-cent moss. I tried a few times, but the hair I produced was uneven in color and texture, and this brought the price down badly. The year my wife and I were married we lived for a while in a little field station that the University of Florida's Biology Department had on the shore of Newnans Lake, a big, shallow cypress-shore lake that used to have a lot of catfish in it. It was depression time then, and our neighbors on the lakeshore were all squatters, who to keep body and soul together trotlined for catfish and soft-shelled turtles; illegally trapped bream, bass, and speckled perch; and picked moss. Marjorie and I weren't in a lot better financial shape than they were, and I used to look enviously at the big heaps of good hair moss our friends would pile onto the truck when it came around. It is years now since I last saw a moss truck broken down on the road shoulder, or even running on the road. In Florida the moss factories have all been shut down, as they probably have been in Louisiana, too. There are trotline fishermen and a few deer's-tongue pickers left in northern Florida, but the moss pickers all seem to be gone. I don't know why, because hair moss was good stuffing—but maybe plastic hair is better. Or possibly picking moss just got too symbolically menial for our time.

Spanish moss should really not be thought of as a plant apart. Its full impact on the landscape comes from its partnership with the live oak tree. Moss and live oaks clearly belong together. On a fair day in a live oak hammock, when the moss and squirrels are fluffy and the tree trunks have lost the black they get from rain, the trunks, moss, and squirrels nosing about the ground seem all so unanimously gray that they must be separate parts of one mutualistic enterprise—which, in a sense, they are. In every way, moss seems most at home in a live oak woods. In the crowns of other kinds of trees with more usual upswept limbs, the moss clumps tend to be bunched in heavy central masses that blot out the light. But the beamlike limbs of live oaks are mechanically stronger than those of other trees and often run out horizontally for thirty or forty feet or more. Deployed along these limbs, the moss drapery looks orderly and purposeful, like a natural appurtenance of the tree. Because it hangs vertically across the horizontal limbs, the internal geometry of a live oak hammock is unlike that of any other forest, and these hammocks seem to me one of the most distinctive landscapes of the Southeast.

In northern Florida two kinds of native oaks have been planted along the streets of towns for a hundred years and more: the live oak and the laurel oak. Spanish moss thrives in both, but the two are very different in their reaction to the moss. In laurel oaks moss shades out foliage and accumulates in masses that, by their weight, break off limbs during heavy rains. I don't know whether moss ever actually kills a laurel oak tree, but something appears to limit the life of the species to less than a hundred years, at least when it is planted on city streets. Although the longevity of live oaks has not been documented in detail, Tony Jensen, Extension forester at the University of Florida—who knows more about the live oak lifespan than anybody I know—has collected evidence that live oaks grow on the average about an inch per year and that they live much longer than a hundred years.[1]

In any case, they live a lot longer than people do. Trees now alive have seen the history of the South unfold. Old live oaks serve as boundary markers in land deeds, and others stand about the country in rows and avenues that tell of vanished mansions and lost roads. There are suicide oaks, and oaks from whose limbs horse thieves were hanged. In the shade of live oaks now alive, charters were granted and treaties were signed or rejected. Sidney Lanier wrote "The Marshes of Glynn" beneath a live oak tree, and, under others, Gabriel was restored to Evangeline, Jean Lafitte plotted his buccaneering escapades, and James Ryder Randall composed his famous song "Maryland, My Maryland." Back in the days before there were any hurricane warning systems, the live oaks along the Gulf coast were sometimes the only stable refuge from the winds, waves, and floods of the West Indian hurricanes, as they called them. Not too many years ago, many of the inhabitants of Cameron, Louisiana, were saved from the fury of Hurricane Audrey by clinging to the live oak trees.

Live oaks live in many kinds of terrain, and from one kind to another they vary markedly in form and size. Besides the big, arched, beam-limbed trees that you see in northern Florida hammocks on well-drained sandy soil near water, live oak occurs as a low shrub on sand dunes and on dry Texas prairies; it mixes with cabbage palms and tropical hardwoods in the little island hammocks of the Everglades; and it joins with swamp bay, red cedar, and dahoon holly on the sea islands of the Gulf coast. In the salt-marsh and Everglades hammocks, the trees form closed canopies only a few feet above your head. On ocean-beach dunes, under the influence of salt spray and a strong prevailing sea wind, they string out downwind, sometimes almost like vines. In northern Florida you sometimes find old live oaks growing in mixed stands with other broadleaf trees, somehow having survived the processes of ecological succession by which a live oak hammock gives way to a diverse moist forest. In these mixed stands the live oaks often have a very different look, with trunks unbranched for thirty or forty feet, or even more, and with upswept limbs and crowns that join the light-sharing mosaic of magnolias, laurel oaks, ashes, maples, and hickories that live there with them.

So the live oak is ecologically very plastic, molding its anatomy to conform to a wide range of living conditions. But when I wax rhapsodic over live oaks, it is not seaside trees, espaliered to the underside of the wind, that I have in mind, or even the soaring oaks of Bear Hammock, before they turned the machinery loose in the legendary Gulf hammock country. It is the moss forests that I think of, where Spanish moss and big, arched, and spreading live oak trees build a fantastic landscape found nowhere else.

The word *hammock*, I ought to stop and say—as applied to landscape—is a Southeastern regionalism used, as far as I know, only in Florida and adjacent parts of Georgia and Alabama. Webster says it is derived from *hummock*, but nobody really knows for sure where the word came from. Actually, it means at least three different things, depending on where you are. In coastal Georgia a hammock is a little island in the salt marsh, usually with red cedar, small live oaks, and saltbush growing on it. Down in the Everglades the term is applied to isolated patches of small broadleaf trees, many of them West Indian species, in the sawgrass or maidencane marsh or limestone pinelands; similar, largely tropical hammocks used to be widespread in the Florida Keys. Throughout the rest of the range of the word, a hammock is any predominantly evergreen woods that is composed of nonconiferous trees. In a country so widely spread with pinelands and cypress swamps the term is useful—or used to be. It is less useful nowadays, because most of the old hammocks are gone.

In the vernacular of the plant ecologists, the moss forest—that is, the woods dominated by the single species, *Quercus virginiana*—is known as xeric (dry) hammock. The term is not a wholly happy choice, but in general, live oak woods are drier than other nonconiferous forests of the region. On ground that in most years is temporarily flooded, an association of bays, black gum, maple, and other water-loving broadleaf trees develops, and this kind of country is known as low, or hydric, hammock. Both xeric and hydric hammocks are considered to be, in the long view, unstable landscapes, which when left to themselves for decades and kept free of fire, grazing cattle, and human ruin are slowly invaded by other kinds of trees. Eventually these newcomers shade out the original forest and form an association in which magnolia, laurel oak, and holly are usually conspicuous.

In northern Florida this is considered the stable end point of the process known as ecological succession. That is to say, in north-central Florida, if geologic conditions and climate stay the same, the magnolia-laurel oak hammock is the "climax" association, the kind of biological community toward which all other types of vegetation—not just hammocks but flatwoods, pine hills, swamps, and marshes—slowly develop. These climax hammocks are mainly evergreen, and though there are numerous deciduous species in them—blue beech, hop hornbeam, cow oak, dogwood, pignut and mockernut hickory, and a lot of others—the year-round shade is evidently deep enough to exclude the seedlings of extraneous species. This is the northern Florida version of the vast original broadleaf forests of eastern North America, and the nearest thing we have to the classic primeval forest with a high closed canopy, a scattering of lower level trees, and a moist, cool, wind-free understory almost wholly devoid of underbrush.

The soil in which these magnolia-laurel oak hammocks grow is good, and only a few of these hammocks have survived. A formerly magnificent example near Gainesville, big enough for two or three Boy Scouts to get lost in every year, is called San Felasco, or simply Big Hammock. This used to be a famous hunting ground, but its turkeys and deer were roosted and fire-hunted out a long time ago. Selective cutting has been done there for decades, with slowly growing intensity. Some of the best magnolias were cut into sections for butcher's blocks; big dogwoods were sawn into billets to make shuttles for English textile mills; people took out little dogwoods, redbuds, and fringe trees to plant in their yards; and recently, more disastrous cutting has been done for pulp and crating mills. But with it all, San Felasco has never been ruined.[2] It is big and wild, and enough old crooked or hollow trees are left to keep some of the original magic of the forest landscape. It is still enchanted, and a great many people go out there just to be in a quiet green place.

There is little moss in such a climax hammock as San Felasco. In the moss forest the trees are mainly live oaks, and because the soil there is generally poor, and because live oak is twisty-grained and hard to split for firewood, and not nowadays used for anything much but platforms for cranes and draglines to stand on when they work in the mud, there is quite a lot of live oak hammock still left in the southeastern coastal plain.

There was a time, however, when it seemed that the moss forests would disappear before any other feature of the coastal landscape and that the only live oak trees left growing there would be on plantations. The wood of the live oak is heavy, hard, and admirably able to withstand the alternate wetting and drying to which ship timbers are subjected. During the latter half of the eighteenth century it became a strategic war material, considered indispensable for ships of the line, frigates, and sloops of war, and even for the new steam vessels they were building. It was too heavy for planking, but it was ideal material for "futtocks and other truncheons . . . beams, stems, sternposts, transoms, breasthooks of upper deck, top timber and low timber." I don't know what all those things are, but they sound vital and make it easy to understand the anxiety of those times over the diminishing supply of live oaks. An American navy of six frigates was authorized in 1794, stimulating frenzied speculation in live oak timber, and by the time the Department of the Navy was established in 1798 live oak was growing dangerously scarce. The shortage moved Congress to approve the purchase of live oak lands in Georgia and South Carolina, and in 1814 more tracts were acquired in Louisiana and Alabama. In Spanish Florida, live oak had for decades been an important export to Europe and the West Indies, and when Florida became a U.S. territory immediate steps were taken to protect the remaining hammocks from timber thieves and smugglers. By 1825 it was estimated that as many as one hundred fifty vessels a year were carrying the timber out of the St. Johns River, and all the groves on the lower reaches of the river had been cut.

In 1832 Audubon traveled up the St. Johns with a government live oak agent. In his essay "The Live-Oakers" he had this to say of the people who cruised the woods in search of trees with the natural crooks that conformed to patterns they carried with them for curved ship timbers:

> The men who are employed in cutting the live-oak, after having discovered a good hummock, build shanties of small logs, to retire to at night, and feed in by day. Their provisions consist of beef, pork, potatoes, biscuit, flour, rice, and fish, together with excellent whiskey. They are mostly hale, strong, and active men, from the eastern parts of the Union, and receive excellent wages, according to their different abilities. Their labors are only of a few months' duration. Such hummocks as are found near navigable streams are first chosen, and when it is absolutely necessary, the timber is sometimes hauled five or six miles to the nearest water-course, where, although it sinks, it can with comparative ease, be shipped to its destination. The best time for cutting the live-oak is considered to be from the first of December to the beginning of March, or while the sap is completely down. When the sap is flowing, the tree is "bloom," and more apt to be "shaken." The white-rot, which occurs so frequently in the live-oak, and is perceptible only by the best judges, consists of round spots, about an inch and a half in diameter, on the outside of the bark, through which, at that spot, a hard stick may be driven several inches, and generally follows the heart up or down the trunk of the tree. So deceiving are these spots and trees to persons unacquainted with this defect, that thousands of trees are cut, and afterwards abandoned. The great number of trees of this sort strewn in the woods would tend to make a stranger believe that there is much more good oak in the country than there really is.

Some of the best live oaks grew along the shores of the remote bays and estuaries of the Florida coast. These were surveyed, and wardens and coastal patrols were detached to guard them; but most of peninsular Florida was too wild to be adequately protected, and the smuggling continued.

When the Pensacola navy yard was established, big tracts of live oak were acquired in neighboring areas in an effort to stabilize the supply of ship timber. Colonel Joseph White, territorial representative, was appointed to clear the titles to this land, some of which was on Santa Rosa Island across the bay from Pensacola. Part of this Santa Rosa tract was by chance owned jointly by Colonel White and his good friend, Federal Judge Henry Marie Breckenridge, a former Pennsylvanian with some botanical knowledge. These two government servants mounted a vigorous lobby for the establishment of a live oak plantation on their Deer Point property. In a long letter Colonel White reminded the Secretary of the Navy that live oak was "the most essential material in ship-building, equal in strength and durability to teak." Later he said, "We shall be in half a century without live oak enough to repair such a vessel as the *Constitution*," and he reminded the Secretary that "the late Emperor Alexander of Russia instructed his agents in this country to procure for him several barrels of acorns from the live oak, that he might have them planted on the southern part of his domain."

The Santa Rosa plantation was eventually approved, and Judge Breckenridge was appointed supervisor of it. He was directed to brush out and prune the natural hammocks, to cut avenues to serve as fire lanes and truck roads, to transplant young trees to more suitable places, and to plant a two-hundred-acre nursery plot with acorns. When Andrew Jackson took office, however, his new Secretary of the Navy, John Branch, was not enthusiastic about the undertaking. He eventually cut off support for it, and the project languished and died. Breckenridge went sadly back to Pennsylvania; but he must be remembered as the first American public forester, and the Deer Point plantation as the first of the country's managed forests.

If the navy had continued to sail wooden ships the live oak shortage might have become disastrous. But the Civil War broke out, and in 1862 the Confederates raised the *Merrimac*—a federal frigate that had been scuttled at Norfolk at the time Virginia seceded—armored her with iron plates, and sent her into Hampton Roads. There she sank a lot of ships in a very little while and might have destroyed the whole Union Navy single-handedly had not the federal ship *Monitor* been hastily built and sent out to stop the mayhem. The fight between these two ironclads revolutionized war at sea and made wooden warships obsolete. The value of live oak timber quickly declined, the cutting stopped, and the moss forests were left to grow again. From then until now they have been less industriously ruined than most other kinds of woods. When the frigate *Constitution* was restored for exhibition a few years ago, the timbers used were taken from a pond at the naval air station at Pensacola. They had been cut on the Santa Rosa reserve before the Civil War and since then had lain there "drowned," as they say of timber stored underwater. That was probably the last time live oak will go into a ship of war.

There was a momentary reawakening of interest in live oak as an article of commerce, however, when the Seminole Indians asked more reimbursement for the land they had ceded to the United States in two treaties, one signed in 1823, at the close of the First Seminole War, and the other, the treaty of Paynes Landing, in 1832. The territory involved was virtually all of Florida south of the old Spanish road between St. Augustine and Pensacola, except for the Indian reservations and some Spanish grant lands. Almost from the outset the Indians complained that they had been cheated, and eventually they took a formal petition to the Indian Claims Commission. In 1966 they were awarded $12 million. Though this was acceptable to the deported section of the tribe now living in Oklahoma, it was turned down as insufficient by the Florida Seminoles. The tribal lawyers have appealed the decision and are asking for an award of about $20 million.

One of the expert witnesses called in to evaluate the assets the United States gained in the deal was George S. Kephart, a forest economist of Silver Spring, Maryland. In his deposition he calculated that on the seller's market of that period the live oak timber alone was worth at least $3.7 million. Adding to this the value of the stands of cypress, pine, and especially red cedar—which, like live oak, was then an irreplaceable material in naval construction—the timber on the ceded lands was worth at least $20 million. Kephart pointed out that previous lower estimates of the worth of the live oak had been based on a claim that most of it grew too far from "quick water"—that is, flowing streams—to be floated out to market. Kephart explained that the presence or absence of quick water was irrelevant in appraising live oak lands because the logs were too heavy to float and had to be carried out on lighters. There was plenty of water navigable by barges in the regions where the best stands occurred.

Although no longer used as timber, live oaks are not wholly immune to the prevalent ruin. Only a few miles from where I live a paper company bulldozed a beautiful hammock a little while ago and planted in its place dismal rows of slash pine seedlings.

These tree farmers may turn out to be one of the most important factors in the loss of original landscape. When you look down into the peninsula of Florida from an airplane, you still see quite a comforting amount of open country, and some people take this to mean that most of the land is wild. This is not true. It is mostly farmland you are looking at, great seas of agriculture lapping at each other's edges and at the edges of the places where people are. These oceans of man-made green are huge citrus groves, planted pasture, sugar cane, and truck farms—if it is the far South you are in—and planted "forests" of slash pine trees. These are mixed blessings. They all are ecologically near the nadir, the merest skeletons of ecosystems. You can walk a mile through any of them and see almost nothing to relieve the eerie absence of extraneous life. Of the four, the planted forests have the edge, ecologically speaking, but it is only a very slight one. And in a way the tree farms are the worst of the lot because they have the superficial form of wildwood and yet have almost nothing of the character, diversity, and complex organization that a real forest has. A pure stand of managed pines is impeccable agriculture and nothing more.

In spite of this, planted forests would be welcome as simply buffers against floods of unnecessary people, if only the public relations men that the industry engages would desist in their campaign to represent tree farming as conservation and tree farmers as powerful, foresighted stewards of the natural world. The necessity for plantations in which a single kind of tree is set out in rows no doubt exists; but to create the illusion that the resulting landscape is a "natural" resource and a wilderness heritage for the public's children is more dangerous than it looks.

In my book *Ulendo: Travels of a Naturalist In and Out of Africa,* I suggested that probably in the long run the most destructive enemy of the natural world will turn out to be the capacity of humans not just to change nature and environment but to be persuaded to like the changes, no matter how dismal they are—just so no obvious public health hazard is involved.[3] The pulp and lumber industries are depending heavily on this. There is mounting wood shortage in the world, and growers are planting oceans of pines, and this is possibly necessary; but instead of just letting it go at that—apologizing and explaining the need for lumber and paper pulp—the companies have set out to delude people into believing that because they plant trees instead of corn they are creating forests. This corrupts the public taste for original wilderness by imbuing the urban population, already badly out of touch with classic natural values and with next to no real understanding of ecology, with a false sense of security over the way the dwindling remnants of the wild world are being cared for. While pushing over a rare sample of original forest to make room for rows of pine trees is sad to see, deluding gullible people into thinking that the pine farm is an "improvement" is downright wicked.

But that is the strategy afoot, and the standard for blatant obfuscation now prevalent among the great planters of trees was set in April 1967 by an advertisement put out by Olin (chemicals, metals, paper and packaging, housing, arms and ammunition). You may recall the page-and-a-half picture of a handsome piece of natural forest and the eye-catching heading: "If You Think It's Beautiful Now, Wait Until We Chop It All Down." How could you forget it! It was obviously the aim of the ad that after getting over your first momentary shock you would decide they didn't mean what they said, that the heading was really just an ingenious—really quite whimsically courageous—tactic to get you ready to receive a solemn moral. But as you read on down through the column you gradually realized that, by God, they really did mean it—not one hundred percent literally maybe, but almost one hundred percent. Very clearly it was pointed out, for instance, that "most virgin forests are already doomed" and are "waging a constant, losing battle against disease, insects, and wildlife," while in an Olin forest "protective sprays and fertilizers are applied," and "the result is a healthier, more abundant forest. One that's far more productive than the original forest." The ad went on, in carefully chosen half-truths, to generate some utter confusion out of which, its creators clearly hoped, there would emerge the impression that the pulpman's values are really the same as those of other passionate stewards of the natural world. And finally, down at the end of the column, the admen asked the reader please to excuse them if they "ignore the dated cry of 'Woodman, spare that tree!'" and then wound up with their own ringing cry: "We're not ashamed of what we do."

Since that advertisement appeared there has been so much of that particular sort of dangerous nonsense that even to call attention to it now may seem superfluous. But at the time it appeared, I was stirred to write to the *Saturday Review*, the magazine I saw it in, and protest its publishing such stuff. What I said was this:

> Today I opened your April 8th issue and saw the Olin advertisement with the picture of the forest and the heading: "If You Think It's Beautiful Now, Wait Until We Chop It All Down." I looked for a humorous twist to the eye-catching statement; but the Olin people were not joking. The advertisement is a grotesque effort to sell the idea that tree farms are not just necessary disruptions of natural landscape, like cornfields and parking lots, but are a substitute for forest, and actually superior to the original thing. Not merely superior in producing pulpwood. Biologically and aesthetically superior. Most perfect, more beautiful, more eternal, one gathers.
>
> I can't make out whether the ad is a cynical attempt to brainwash the public or is just honest ignorance of an unbelievably puerile sort. But in either case the main blame lies not with the exploiter straining for self justification but with *Saturday Review* for selling space

to purveyors of irresponsible deception. I am writing to tell you how disappointed I am to see this strange breakdown in the judgment of a magazine I have admired.

Obviously your science editor never read the Olin ad. But it is not just principles of ecology and wildlife management that are violated in it. Olin is evidently out to wreck one of the humanities: the intelligent appreciation of original, naturally organized nature. For a magazine dedicated to the humanities, *Saturday Review*'s connivance in the messy campaign is a sorry thing to see.

In a gracious answer to my letter the *Saturday Review* editor explained that the advertising and editorial departments of that magazine have a long-standing agreement not to tell each other what to do. I was sorry to hear that. I hope they have worked out a better agreement by now. Because espousing propaganda like that Olin ad cancels much of the good the magazine has done through its editorial policy of spreading word of the loss of the natural world.

As habitat for animals the moss forest is not populous. It gets too dry, and there is not much for a year-round resident to eat. The main food resources are the thin layer of humus, built up by decomposition of the penny-hard little live oak leaves, and the unpredictable crop of acorns. Live oak acorns are low in tannin and less bitter than those of other southeastern oaks. On some trees they seem almost worth eating. The Indians regarded them as a staple resource of the region and made cakes of acorn flour or used it to thicken stews. Some years the acorns fall in profusion; some years there are almost none. In the good years gray squirrels, wild turkeys, and deer rejoice, wild hogs get fat, and our eighteen-year-old Guernsey cow spends hours every day nosing along the edges of the pasture, picking up acorns one by one. As wide-mouthed and stiff-lipped as the cow is, it seems strange to me that she can pick up the little smooth acorns in worthwhile quantities. I suppose the answer is motivation. That cow has made milk for fourteen calves and five children, and she did it all on northern Florida pasture, where after the first freeze the Bahia grass is only roughage, and the scant green stuff in the woodland range is quickly browsed away. Under stress like that it is sink or swim for a poor old cow, and what ours did was learn to pick up acorns.

The most spectacular event of the big acorn years is the coming of hosts of wood ducks to the live oak woods. One January morning, for example, when the acorn crop was prodigious, we saw a great patch of ground alive with a silent host of wood ducks. Where an hour before only squirrels had been puttering, a flood of ducks now moved across the ground beneath the oak trees, industriously stuffing themselves with acorns. The following year, again, it was a good acorn season, and once more the wood ducks came to the hammocks. They arrived in flocks of half a dozen or twenty or thirty, circled low over the woods, communing in a quiet, whispering whistle that I can't remember having heard from wood ducks before, and then either pitched on a little sinkhole pond at the edge of the pasture and walked out under the live oak trees or dived directly down into openings among the trees. I would give a lot to know how these ducks learn the location of the separate acorn trees and what they do in the poor mast years.

Except during acorn time, the fauna of live oak woods is meager. After the squirrels have searched out or hopelessly lost the last of the season's acorns, the interior of a big hammock is a quiet place, and most animals of any size that one meets there will likely have come in from somewhere else. If you hear a clamor of birds, for example, it will no doubt be a posse of blue jays, cardinals, and wrens that have come together to berate a rat snake one of them discovered. The rat snake himself may have spent the day before in a fence row a half mile away. I have no proof of it, but I think rat snakes travel a long way in the foraging. Several times I have felt sure I recognized one that I had previously seen in some distant part of the farm. Long ago I ought to have started tagging rat snakes that have turned up around our house, but I actually got around to marking only one—or rather I recorded the peculiarities of his pattern for later recognition. This was a slim, energetic snake that stood off the dachshund that found him by coiling in a high spiral and bravely striking at the snapping dog, finally snagging the dog's nose with a tooth or two and sending him shrieking away. I caught the snake, and after recording the distinctive features of his pattern, carried him almost a quarter of a mile away and let him go in a patch of gallberry bushes. That was at nine in the morning. At four that afternoon the snake was back in the yard, not thirty feet from where I had found him before. I won't suggest that he was looking for the dog. I do believe that he did a piece of purposeful pathfinding back to a place he wanted to be in, however, and his journey seems to support the notion that rat snakes are wide-rangers.

If instead of a snake it is a barred owl that you find birds scolding in the woods, the owl may sleep through the following day two miles away in Wacahoota Hammock. In some seasons flying squirrels squeak in the moss forest in the evening; and if you have the patience to sit long and quietly enough at twilight you may see one glide across the space between two trees. But then at other times the flying squirrels are quiet for weeks or months on end and appear to have left the live oak woods completely, though I don't know why or where they go. You may encounter a possum if you walk quietly in the hammock, or an armadillo snuffling and grubbing about, his face up to the eyes in the leaf mold. Both of these are ecologically open-minded animals, which in northern Florida walk on any ground not under asphalt.

The more permanent ground-dwelling inhabitants of the hammock are mostly small and retiring and are closely linked in their feeding to the thin leaf mold and its inhabitants. Besides a host of nearly microscopic arthropods, there are half a dozen kinds of cursorial wolf spiders, a number of crickets, some millipedes and little snails, the spadefoot toad, and the diminutive ground skink. If you roll dead logs over or spend a morning raking trash away from the bases of trees you will rout out some big, black, shiny beetles named *Passalus cornutus;* a few wireworms, which are the larval stage of click beetles; a centipede or two; a couple of kinds of newts; a short-tailed shrew; and any of several kinds of little burrowing or litter-inhabiting snakes.

One of the snakes you most often catch by raking leaves in a hammock is the coral snake. The best place to look for one is not in the depths of the forest, however, but down where it begins to merge—as the best live oak hammocks usually do—with the shore vegetation of some body of water. Even in the very best coral snake habitat you can walk a long way without seeing one out in the open. Most of the time they squander their candy-striped decor down out of sight beneath the oak leaves.

But one ought not to generalize about the habits of coral snakes. They usually shun the limelight, all right, but they have a venturesome streak about them all the same—a talent for the unexpected. I once found a coral snake high in a tree, for instance—twenty feet up in the top of a cabbage palm behind our house. It was a summer evening, and from inside the house I heard a thrashing sound up in the crown of the palm. Thinking it must be a rat snake raiding a redbird nest, I got a ladder and climbed up to see what was going on. Sure enough, the first thing I saw was the flailing tail of a striped rat snake. But looking closer, I could see that the rat snake was not eating any little redbirds. It was being swallowed by a big coral snake. That was for me a wholly unexpected view of coral snake personality.

What I think of as the most quintessential denizen of the moss forest is the little big-eyed, solitary spadefoot toad. As in the case of coral snakes—and for the same reason—many people spend years in Florida without ever seeing this engagingly eccentric animal. To find a spadefoot, you ought to go out into the woods at night with a tightly focused headlight and play the beam slowly around over the oak leaves on the ground. If it is a good hammock, and a good night, you will almost at once see a lot of tiny lights that sparkle fiercely like cut jewels or little stars. Those are the eyes of spiders. More sparkling may come from the trunks of some of the trees, and that is made by spiders too, big spraddle-legged tree-trunk spiders of the genus *Dolomedes;* or, if the light on a trunk has a rosy color to it, it will be a resting moth that you see. Off at greater distance the beam of the flashlight may strike a fire of many times more candlepower than that from the spider's eyes, and this at first looks like a giant spider; but then it turns or bobs about, or moves off fast across the ground, or climbs a hanging grapevine; and this is bound to be a possum. There are other little lights of a wholly different kind, without any sparkle at all but only a glow—not point sources, such as stars seem, but pinkish disks, like tiny paper lanterns or little moons. If one of these glides away across the ground, or just suddenly goes out, it will be a rabbit that is there, a swamp rabbit up from the lakeshore, or a cottontail from a nearby fence row, out in the hammock after Lord knows what. But if such a pink, moonlike light holds steady in your beam, not moving about but only blinking a bit, or slightly shifting its position—if it is low to the ground and little, and is obviously the eye of some calm, contemplative animal—you will have come upon one of the admirable small creatures of the Earth: the mild, pacific, self-reliant spadefoot toad.

Spadefoots are not by any means confined to live oak hammock. Like most of the other inhabitants there, they are found in other kinds of country too, being more concerned with parochial features of their environment—the microhabitat, as it were—than with the kind of landscape in which it occurs. What a spadefoot needs is clean hopping ground for his modest hunting forays, a supply of little animals to hunt or take by patient ambush, and friable soil in which to dig the shallow burrow that he lives in. These conditions prevail in hammocks, where the shade of the trees keeps the ground free of heavy undergrowth. The little, hard fallen leaves arrange themselves in a fairly smooth mosaic that a short-legged toad can easily hop across. The litter and mold they make is shallow, too, so the sand is not too far down for a toad to dig a burrow in. The spadefoot does his digging with a horny spur on the back of each hind foot. The burrow is a short, slanting tunnel, only a few inches deep and just big enough to back into and stay in till dark and come out of again if the night should be one that pleases him.

When Paul Pearson, now president of Miami University at Oxford, was a graduate student at the University of Florida, he studied the population ecology of the spadefoots of the region. His marking experiments revealed that in the best of his spadefoot hammocks there were 2,503 toads per acre. To me that seemed an astonishing lot; but I was even more surprised to learn how much time the spadefoots spent in their burrows without emerging. On the average they came out only 29 nights a year. The average toad spent 9.45 consecutive nights underground, and the most withdrawn stayed below for as long as 104 days at a time. It is hard for me to see how such a conservative predator could get enough hunting done to support life. Perhaps some of the toads counted as nonemergers—simply because Paul never shined their eyes—actually spent the night just inside the mouths of their burrows, out of range of the collecting light but in a position to pick up the occasional insects that strayed within reach. But even so, huddled back into their moist, dark little caves, heads bent down, eyes closed, legs and arms folded in, Paul's spadefoots seemed very introspective.

Even when spadefoots come out of their holes they usually don't go very far. Paul Pearson found that his toads moved, on the average, only 3.66 feet away from their burrows during a night's foraging. Although most of them kept the same burrow throughout the time of the study, thirty-one changed burrows, never to return to the previous one, while eight alternately occupied two, three, four, or five different holes. It is hard to say what all this means, but it certainly sets one pondering.

That is not the end of the engrossing traits of spadefoots, either. Their breeding habits are also strange. Like most other tailless amphibians, a spadefoot must go to water to reproduce. Most other species must move off to the nearest pond or ditch to lay and fertilize their eggs when rain comes at the proper time of year. But spadefoots never join the other kinds in celebrating any ordinary spring rain. It takes violent weather to turn a spadefoot's mind to sex. What he really needs to awaken fierce lust is a howling tempest, with a couple of inches of sudden rain falling in a few hours' time. This is no light fancy, either, because if no big storm comes, the toads simply refuse to breed, staying out in the forest in unfruitful celibacy for whatever period may pass without a properly erotic cloudburst. And what seems every bit as odd as this insistence on rough weather for breeding is the physiologic readiness of the population to breed at whatever moment a storm may arrive. With almost no notice at all, spadefoots can mobilize ripe eggs and sperm and the fervor to hop for a quarter of a mile across country to the place where a few local males are singing about a good storm pool they have found.

This stubborn tendency to bide his time and be stirred out of the comfort of a woodland burrow only when a frog-choker of a storm arrives can be taken as a sign that the spadefoot toad has had a racial history of desert living. In deserts the only frog equipped to reproduce—by the classic amphibian plan, at least—is one constantly ready to take advantage of the sporadic flash floods that come in that kind of country. And, as a matter of fact, some of the living relatives of the eastern spadefoot are desert species. In their case, the adaptive value of such opportunistic breeding seems clear, because the storms bring the only water they ever see. As to why *Scaphiopus holbrooki,* a creature of the moist, forested, eastern United States, should keep up the curious tradition, some other explanation must be looked for. One such would be that Florida toads are just conservative and, like many people, are on general principles loath to give up ancestral ways. But in biology this kind of interpretation is risky. Few animals cling to a complex pattern of behavior if there is no strong advantage in it. And in the present case there would appear to be obvious advantage: the immunity to predation that the habit of breeding in storm pools affords.

Any frog that waits until there is a stump-floating storm to breed, and then goes to the new rain pools to lay its eggs, will automatically be selecting for its young a secure nursery in which there has not been time for a fauna to develop and where predators are consequently few. Tadpoles of pond-breeding frogs are beset by a host of enemies. But the queer sex habits of spadefoots cause their eggs and young to be lodged in pools so lately formed that they have not been colonized by even such pioneer predators as the carnivorous water bugs and beetles, whose own rapacious larvae are aquatic but which as adults fly all about the countryside. The obvious drawback to this plan is the likelihood that the pool will dry up soon after the storm that made it has passed. The spadefoot circumvents this hazard by simply speeding up its incubation and larval development. It is not known exactly what the accelerating factor is, but there is a clear response to the shrinking of the pool as the water evaporates or drains away, and spadefoots have been known to go through their whole development—eggs, embryos, and tadpoles—in six days' time.

To look at Spanish moss hanging all about a hammock in soft clumps and curtains, one might think it would be a great asset to the inhabitants of the place as material for nests. But not many animals use moss this way. They probably avoid it for the same reason that people camping out do—because the covering of little scales holds a lot of water, and the moss gets clammy under the weight of your body. Cured moss makes a good bed, but a live-moss mattress is atrocious. When I was a Boy Scout, air mattresses had not become prevalent, and moreover, it was the feeling then that campouts should prepare Scouts for rugged, virile living later on. I remember how on nearly every camp some tenderfoot would hopefully start pulling down moss for his bed and would be warned by old hands to cut it out. "The stuff is full of ticks and chiggers," the seasoned men would say. " They'll eat you alive." That was not true, but you used to hear it a lot. I never saw a tick in Spanish moss that had been hauled down out of a tree, or a chigger, either. Personally, I doubt that ticks and chiggers have the continuity of purpose to climb a tree. What I decided was that campers, kept awake by the cold feel of mashed live moss, were just sensitized to the bites of the redbugs they had picked up during the day.

Gray squirrels live intimately with moss; but in building their nests it is mainly leaves they use, often mixed with fiber from the leaf bases of the cabbage palm. If any moss is used it is hair moss, put in mainly to tie together the mass of leaves and palm fiber. The only bird that habitually nests in living clumps of moss is the parula warbler, which in the deep Southeast apparently never builds its nests anywhere else. In the North the resident race of the same bird chooses festoons of the pendulous lichen usnea to nest in—which seems an oddly consistent attachment for fuzzy gray epiphytes as a nesting medium.

The only other vertebrates known to bear and raise their young in bunches of Spanish moss are three kinds of bats: the eastern yellow bat, the Seminole bat, and the eastern pipistrelle. When William Jennings, the rabies expert, was studying the ecology of Florida bats as his Ph.D. research, he went to the moss factories and offered a bounty for any bats that came in on their moss trucks. He got 296 bats this way; 179 of them old enough to fly and 117 nonflying young. I was glad to hear about this, because it seemed to explain why rat snakes so frequently climb about in live oak trees at times other than bird nesting season. I mentioned the rat snake as one of the few larger vertebrate animals that turn up consistently in a live oak hammock. I have often seen them inching quietly along a limb, or threading their way through the moss, and used to think they were up there after flying squirrels housed in the hollows. A few times I have even found rat snakes coiled in moss, and I figured these were probably resting there to elude the sharp-eyed snake-baiting birds. But perhaps it is bats that these snakes are after. The species is well known as an avid hunter of bats. It is often found among the rafters of old buildings that have bats in them, and in Florida, in any cave in which bats aggregate, rat snakes are likely to gather about the entrance or even to go some distance back into the dark interior.

For months after the Spanish moss got sick there was no sign of live moss around our house or anywhere along the road into the university or in the old trees on the campus. The corpses of the plants remained, but they hung in dismal, brown, weightless stalactites that blew out level with the slightest breeze and looked like strands of frayed rope when it rained. The second summer of the epidemic I went away and stayed for nearly three months. When I got back home and the car stopped in the yard, I anxiously looked up into the trees to see how the moss was. At first it seemed just the same, or possibly a little worse. But then I got out of the car and walked over to a tree and examined a hanging moss rope; and all along its length there were slivers of silver where seeds had sprouted and inch-long seedlings had grown while I was away. It was a pleasant thing to see on coming home, and I quickly telephoned Tony Jensen to ask whether the moss was reseeding in other places too. He told me that all through the region of the epidemic, new seedlings were growing in the dead clumps, and that in a few places a whole new drapery had developed. He said the most rapid regeneration was in places exposed to long periods of full sunlight.

I thanked Tony and went straight out behind the house to where a pear tree grows beside our garbage pit. I went there because moss used to thrive intemperately in that particular tree. And now, sure enough, on a limb on the southwest side of the pear tree there was a whole clump of brand new moss. It spread for a foot along the limb and hung in short, healthy spikes that had the frosty silver look of the good days before 1968. I was as pleased as if I had come upon a panther holed up in a blowdown out there—more so, because the panther would surely have gone away, while the moss might be coming back for good.

The fast regeneration in strong sunlight seemed to support the notion that a fungus was the cause of the disease, but I reserved judgment on the matter and still do. The way the world is being abused, it will take clear proof to convince me that the exudations of civilization were not involved in the epidemic—if not as the actual agent, then at least by weakening the moss and making it susceptible to disease.

Anyway, I stood there for a while looking at the strong young moss in the pear tree, meditating on the appeal this curious upside-down plant has always had for me; and my thoughts took me back to an April night that seemed to hold it all. It was a night in the spring before the moss got sick. The moon was high and almost full, and a wind aloft was trundling a train of little clouds across the sky. Down in the woods there was no breeze at all, and the air there was so cool and quiet it seemed you could hear for miles. I had gone out to listen to the night sounds of April and to see the almost unreal look of a moss-draped live oak hammock in strong moonlight. The vesper mosquitoes had gone away to wherever they pass the night. The moss hung like spun silver, fluffy, weightless, and stark-still in the motionless air. I leaned against a tree and waited, and a barred owl called on the other side of the pond, and another answered away over in Marion County. It was too early for the summer frogs to sing, but a single spring peeper, left over from the winter choruses, whistled his lonely note in the black-gum swamp at Smoky Hollow. Almost too far to hear, even in the quiet of such a night as that, the four-note call of a great horned owl drifted in across the treetops from the edge of Tuscawilla Prairie, and after that there was only the moss, filtered moonlight, and a quiet so deep you could hear a wolf spider charge a cricket across the dry oak leaves. The rustle of a spadefoot foraging was like a buck in bushes; the repeated squeak of a flying squirrel, in lingering concern over the barred owl's bellow, seemed a strident noise. The moon blazed and faded, I remember, as separate high clouds went by on a wind too high to stir the moss; and one moment the vaulted rooms of the moss forest were flooded with silver light, and the next the glowworms down at the pond edge were torch-bright in the dark.

In Praise of Snakes

When Rachel Carson chose the name *Silent Spring* for her epoch-making book, the silence she had in mind was lost birdsong on a poisoned Earth. The book was a powerful document, and people took heed of it. Birds became not just objects of concern but a symbol of our predicament. This was good for the birds, but it left unattended a lot of other creatures that had no songs to start with and had been silent all the time.

Snakes, for instance. I want to speak in behalf of snakes. I live in snake country, and have always liked snakes and have kept them steadily on my mind. There is a dearth of good census data to prove it, but snakes seem to me to be disappearing very fast. Their survival problems are much worse than those of birds. Besides the waning of their food supply as rodents and frogs grow fewer, and besides the growing toll that cars take on multiplying highways, snakes face the indifference or active antipathy of most of the human race.

People who resent snakes explain their attitude by saying that some snakes are venomous. But there is more to the hangup than that, a lot more. For example, if you pull an angry wildcat out of a bag the reaction of witnesses is no more than simple alarm, or perhaps even grudging admiration. But take out a *harmless* snake and the faces of the people suggest that you have somehow flouted common decency.

I once put into print the risky proposition that a lingering uneasiness over snakes may have come to us genetically from forebears prone to be eaten by pythons and the like. I don't fight for the notion, but I can't logically discard it. In any case, those old days when a proper fear of snakes was the mark of the successful ape are long gone, and the remaining vestige, if it exists, is just a useless atavism, and people ought to dominate it, as they dominate the dread of high places or of eating octopus. Above

From *Audubon* 73, no. 4 (1971). Copyright © 1971 by the National Audubon Society, 700 Broadway, New York, N.Y. 10003.

all, we ought to keep people from conditioning it back into our young by shrieking when a snake shows up.

I am not suggesting that anybody take asps playfully to their bosom or wrestle anacondas or even walk heedlessly through a palmetto flatwoods in Florida. But the United States is blessed with a great diversity of harmless snakes, the numerous members of the huge family *Colubridae* that do no damage to anyone. Actually, the negative ring of the word harmless is inappropriate for these creatures, because they have far more positive virtues. Besides being ecologically useful, they are artistically decorated, richly colored, and consummately athletic beings that really ought to be appreciated more. They ought to be watched. People watch birds regularly nowadays, but you rarely come upon anybody watching snakes. That is a great shame, because snakewatching, though a more demanding exercise than birdwatching, can be every bit as rewarding.

Snakes are harder to find than birds. They don't fly about or sing or do a lot of overt things that help the would-be watcher find them. And even after you have located a snake, purged yourself of any remaining scrap of prejudice, and made ready to witness engrossing natural history, it may be quite a while before the snake does anything. Birds of course fidget constantly, except perhaps owls and fishing herons. There is never a dull moment in their company. Snakes, on the other hand, being cold-blooded, live more deliberately; and if you have the misfortune to come upon one that has just eaten, there may be little or no action for quite a while—for up to a week, perhaps, or in extreme cases even longer. Some people don't have that kind of time to put into the venture and go back to watching birds.

But although snakes are slow in digesting, they are also often slow in coming upon anything to eat, and they spend a great deal of time foraging. It is then that you see them do exciting things.

To court the quiet pleasures of snakewatching, go out to the edge of town and walk along a grown-up fence row till you come upon a racer, say—a blue racer if it's the North you are in, or a black snake if you are in a Southern state. The racers are subspecies of *Coluber constrictor*. Despite their specific name, they don't constrict, but they do a great many other interesting things and are worth some patient attention. Finding one of them on the sunny side of a fence row on a fair May morning is no problem. The trick is to keep him in view without breaking into his natural routine. If he sees you first he will either bolt and disappear or rear back, coil, and rattle the leaves with the vibrating tip of his tail and then quickly retreat if he fails to scare you. But once your snakewatcher's eye has been sharpened you will begin seeing racers before they know you are there, and then you will be able to watch one thread his way silently through the leaf drift, easing along in slow undulations or creeping on tilting belly scales as he explores with his flicking tongue-tips every old log or pile of debris where a mouse or frog or smaller reptiles might be hiding. If he comes to a pond he will prospect carefully around the edges, moving into and out of the shallows, swimming the small embayments, sliding into low willow to see what bird nests or tree frogs may be up there. To keep contact with a blue racer through a morning's foraging is a challenge that will call out all one's self-discipline. Bird glasses can be helpful, but the main requirement is a serpentine sort of stealth.

A birdwatcher is helped by the singing of the objects of his search. Snakes never sing. Some rural Nicaraguans say that the fer-de-lance is able to whistle, but they are mistaken. Rattlesnakes of course are commonly heard before they are seen, and sometimes one of them will jump the gun and reveal himself by rattling at a man or dog from a long way off. These are probably just high-strung individuals, however, because most rattlers stay quiet until they feel they are about to be stepped on. Another snake that is sometimes heard before being seen is the pine snake or gopher snake (genus *Pituophis*), which rears up in high coils in an impressive way when disturbed, reinforcing the menace with a hissing that must be the ultimate in serpent vocalization. The sound is so loud that you can hear it across an acre of ground, and so stertorous that one subspecies evoked the name bull snake from the early settlers. Pine snakes are active, intelligent animals, really, and quite affable once they shed their racial prejudices; but when first encountered in the woods the show they put on is unnerving, and they sometimes begin their bellowing at you from quite a distance.

Often a mixed chorus of indignant-sounding birds—buzzing wrens, shrieking jays, and chipping cardinals, for instance—will lead you to a snake. Another sign of a snake is a shrieking frog. Some of the most arresting snake behavior is related to their feeding. Watching this is sometimes not for the overly impressionable person, simply because the diets of snakes tend to be heavy in whole, live vertebrate animals. So if you take the second grade out on a field trip and come upon a milk snake engulfing a mouse, or a bullfrog screaming heartrendingly because a garter snake has hold of his leg, some calming apologia may be in order. You might explain that snakes—having no hands, only little prickly teeth, very distensible jaws, and a liking for extremely fresh food—eat mainly whole, live animals. It is simply their custom, you can say, and not much worse—is it, really?—than a man eating a raw oyster or dropping a lobster into a pot of hot water. In any case, a shrieking frog is an almost sure sign that a snake is there, and you only have to make a quiet approach to be able to get close to him.

A more ingratiating bit of snake behavior to watch is the bluffing and playing dead of the hog-nosed snake, or spreading adder. Most people don't live in spreading adder country, but those who do are neglecting a local asset if they fail to make an effort to see the antics of one of these short, corpulent, and inherently amiable serpents when he tries to discourage interlopers by pretending to be first a deadly viper of some kind and then the mere corpse of a snake.

Most people who come upon hog-nosed snakes either conclude that they are venomous and dispatch them or go quickly away. To see their act at its best you should do neither, but behave as follows: first of all, make sure it is indeed a hog-nosed snake you have at hand and not a ground rattler, which would prove to have almost none of the spreading adder's winning ways. Then, move straight up to the snake and sit down on the ground in front of him. He will coil in a purposeful way, rear back and spread the whole first third of his body as thin as your belt, and lunge out at you repeatedly, each time hissing with almost intolerable menace. If instead of recoiling you steel yourself and reach over and pat the snake on the back, his menace will wilt before your eyes, and he will proceed to prove that you have killed him. He will turn over onto his back, open his mouth, extrude his tongue and rectum, and then, after writhing about until his moist parts are all coated with debris, lie there belly-up as clearly defunct as any snake could be.

But don't feel badly about him. Give him two minutes, say, and the catalepsy will wane. He will draw his tongue back in and ever so slowly turn and raise his head to see whether you are still there. Move your hand quickly before him, and he will flip back over into his supine seizure. Reach down and turn him right side up, and he will instantly twist over onto his back again. But then get up and move off a little way and wait patiently behind a tree, and you can watch him slowly come back to life, turn right side up, and quietly ease away.

Another overt piece of snake behavior that once in a blue moon rewards a lucky snakewatcher is the combat dance that the males of some species indulge in. This is a stereotyped, balletlike posturing by two male snakes, a crossing and recrossing of necks or intertwining of bodies that takes place as the two performers raise the foreparts of their bodies high above the ground until they topple over backward. The routine is punctuated by periods in which the snakes chase each other about at terrific speed. Then the neck-crossing is resumed. No biting or constricting or other violence is involved, and the social function of the dance is not clear. Over and above its interest as an arresting instinctive behavior pattern, the occurrence of the ritual in such distantly related animals as pit vipers, rat snakes, and racers is a thing to wonder about. Either the dance has come down to them from a very distant common ancestor, or it has been hit upon independently by all these different modern snakes. Either explanation would seem pretty extraordinary. As simply a thing to see, however, the dance is one of the rare rewards of the fortunate snakewatcher.

If an aspirant snakewatcher has trouble getting out into open country he can find a lot of gratification watching well-adjusted captive snakes. Not captive in the sense of caged, but domesticated, living unrestrained in the house with the dog and children. Every kind of snake is not appropriate for this kind of relationship. Water snakes, for example, get badly underfoot and may bite when you trip over them; racers upset bottles in cupboards; rat snakes insinuate themselves into all sorts of small cracks and secret places, and often get mashed in doorjambs. None of these are insurmountable objections, but the point is that there are better snakes for the purpose; and the best of all, in my opinion, is the indigo snake *Drymarchon corais*. The indigo is handsome and extroverted, and for some ironic reason a natural-born friend to man. It rarely bites even the first hand that lays hold of it in the wild and quickly adjusts to life in houses with people. Not being a constrictor, it never creeps surreptitiously about the walls and woodwork, and being heavier and less nervous than racers, it is not so inclined to whip petulantly about among human legs in a room.

Actually the indigo snake is becoming dangerously rare throughout most of its range, and a person probably ought to go to jail for catching one. Indigos can be bred in captivity, however, and it really ought to be done, because a more ingratiating house pet is hard to imagine. I have known several people who have shared their homes with indigo snakes.

My friends Bill and Joan Partington had a rewarding relationship with an indigo snake that lived in their home for many years. It stayed so long that Joan had a baby while it was there. When she nursed the child the snake would come out of the bookcase it lived in and rest the forepart of its body on the arm of her chair to watch the undertaking. The Partingtons could never be sure whether it was the baby that evoked the rapt attention, the way Joan was built, or both. Bill even suggested that the snake might have thought the child was trying to swallow Joan, but that seems unlikely to me.

Although keeping snakes in homes and zoos is fun, it does not discharge our obligation to save them as wild species in the world around us. Most sensible people have now worked up a healthy fright over the future the world holds for their children, and birds are often drawn in under the umbrella of their concern. But I have heard little worrying over the future of snakes, and this to me is depressing. Snakes are not degenerate beings, punished with leglessness for ancient sins, as people once said. A snake is the elegant product of a hundred million years of natural selection. Its loss of legs was an evolutionary advance, a means of living successfully in unexploited ways. But because those ways are secret, the decline of snakes in our changing world has gone on almost unmonitored. Others have spoken for cranes and whales, and I hasten to say these words in praise of snakes, whose silent spring is also far along.

The Landscapes of Florida

There are no mountains in Florida and only a few rolling hills in the northern part of the peninsula. The rainfall over the state is fairly uniform, and the temperature is moderate, though in most years frost occurs in the northern section. In spite of this uniformity, there are myriad forms of landscapes in Florida. The major factor molding the scenery is its geological history. Several times peninsular Florida has been lifted above the sea and then submerged, causing widespread deposits of porous limestone (which make possible the famous Floridan Aquifer), as well as clay lenses, old dunes, and extensive deposits of sand. The action of numerous springs, rivers, and lakes on the limestone and sand base has resulted in a wide and fascinating variety of plant associations. A familiarity with these different forms of landscape will add immeasurably to the enjoyment of the natural Florida.

Terrestrial Situations

Pine Flatwoods

The flatwoods may be divided into several fairly distinct types, most of which intergrade or interdigitate at points of topographic or edaphic change. In the inland flatwoods the slash pine (*Pinus elliotti*) is typically the dominant tree, although the long leaf (*Pinus palustris*) often mingles with the slash pine where the soil is well drained. Occasionally long leaf pine occurs in pure stands over large areas of level country. In extreme southern Florida, and along both coasts, slash pine and long leaf are replaced by the South Florida slash pine (*P. elliotti densa*). In the coastal strips the size of the trees and the nature of the substratum are essentially the same

Adapted from *A Contribution to the Herpetology of Florida,* Biological Science Series, no. 3 (Gainesville: University of Florida, 1940).

as in the interior flatwoods. In the limestone ridges in the Everglades and on Big Pine Key, however, the trees are widely spaced and scraggly, and the irregularly eroded rock outcrops support little undergrowth. From the standpoint of faunal distribution there appears to be little basis for distinguishing between coastal and inland flatwoods. The significant types are as follows.

WIRE GRASS FLATWOODS. The dominant tree may be long leaf, slash, or South Florida slash pine; the substratum is chiefly composed of wire grasses, of which *Aristida stricta* is one of the most abundant. The soil is sandy and generally poor; it is underlain by a clay subsoil of variable permeability. In low areas the soil may be almost permanently wet, and crayfish burrows are numerous. Shallow ponds, grown over with cypress or maidencane and ringed by a marginal growth of pond pine (*Pinus serotina*) or black gum (*Nyssa biflora*) are frequent. Where regular burning occurs the fauna is very meager, but fire-free areas are usually well populated.

PALMETTO FLATWOODS. The most important difference between palmetto and wire grass flatwoods lies in the nature of the substratum. The characteristic shrubs are saw palmetto (*Serenoa repens*) and gall berry (*Ilex glabra*). The hard-pan is usually well developed, close to the surface, and exceedingly impermeable; consequently the groundwater is not available to the surface soil, and in times of prolonged drought such flatwoods are very arid. In wet weather, however, rainwater and runoff from higher areas may remain in the ground for long periods. Neither the gopher tortoise nor the pocket gopher (*Geomys pinetis*) are found in typical palmetto flatwoods.

LIMESTONE FLATWOODS. This unique association is developed in the Miami Oolite, which extends southward from Delray in Palm Beach County to Homestead in Dade County, whence it continues westward as a series of insular ridges that form the southern boundary of the Everglades for about two-thirds the distance across the tip of the peninsula. Soil is scarce or practically lacking over most of the region; the oolitic limestone is very porous and is pitted and honeycombed with solution holes of various sizes. The dominant tree is the South Florida slash pine; the ground vegetation is composed of a number of xerophytic shrubs and herbs, most conspicuous among which are wire grasses (*Aristida Spp.*), saw palmetto, silver palm (*Coccothrinax argentea*), and coontie (*Zamia floridana*).

Upland Forests

ROSEMARY SCRUB. Scrub is found in patches all over Florida, but it covers extensive areas nowhere except along the lower east coast and in the central Lake Region. It is essentially an old dune association; the soil is almost pure white (St. Lucie) or yellow (Lakewood) sand, in some places forty or more feet deep. Regardless of the color of the deeper parts, the thin bleached surface layer is usually gleaming white. The only tree of any size is the sand pine (*Pinus clausa*). The undergrowth consists mostly of xeric woody shrubs or dwarf trees, including sand live oak (*Quercus geminata*), myrtle oak (*Q. myrtifolia*), rosemary (*Ceratiola ericoides*), saw palmetto, and several ericaceous species. In recently burned areas the shrubs form dense, thicketlike tangles, but where the pines are mature and closely spaced the lower levels of the forest are more open. In such parklike groves, locally known as strands, the glaring white sand is partially concealed by several species of lichens ("reindeer moss"), and rosemary often grows in scattered clumps. Rainwater sinks almost immediately into the loose sand. In dry weather the evaporation rate is very high, and the temperature fluctuates rapidly and reaches considerable extremes. On several occasions at sunrise I have noted temperatures lower by ten degrees or more in the Marion County scrub than in the cut-over pine hills six or eight miles away. Deer, wildcats, skunks, and foxes are fairly common, while the grasshoppers *Schistocerca ceratiola, Melanophus forcipatus, M. indicifer*, and *M. tequestae*, the spider *Lycosa ceratiola*, the Florida jay *Aphelocoma caerulescens*, and the lizard *Sceloporus woodi*, seem to be entirely confined to the scrub.

HIGH PINE. Once widely distributed throughout Florida, the high pine has mostly been destroyed by lumbering and agricultural activities. The principal tree in this association, the long leaf pine, is a highly valued timber tree, and the soil of the pine hills, though sandy and rather dry, is much more suitable for cultivation than are the flatwoods soils. The topography is rolling, and the water table is generally very low. The chief lower-level plants are blue jack oak (*Quercus incana*), turkey oak (*Q. laevis*), twin oak, myrtle oak, saw palmetto, and wire grass. Cut-over hills usually support an extensive second-growth of blue jack or turkey oak. Pocket gopher and gopher tortoise burrows are usually abundant.

Hammocks

In Florida the word *hammock* is applied to any hardwood forest. The prevalence of coniferous woods—pinelands and cypress swamps—lends significance to the term, which distinguishes between these common types and the hardwoods. Hammock soil in general is the most fertile in the state; humidity is higher than in high pine and less fluctuating than in flatwoods; a well-developed humus layer usually covers the soil. Most of the many hammock types intergrade with one another and with other associations. In addition to whatever special food relationships may influence the character of the hammock faunas, it seems probable that the most important ecologic factors are the presence of leaf mold and the usually lower and more constant evaporation rate. Because almost no reptile or amphibian seems to be restricted to any single type of hammock, it is only necessary to distinguish between the three following major types.

LOW HAMMOCK. This is a loose term applied to woods growing on low, damp, wet, or flooded ground, and it includes conditions intermediate between mesophytic hammock and cypress swamp. The trees may be nearly all of one or two species—as in the cabbage palm–red bay hammocks of the peninsula and the gum swamps of North Florida—or a variable association that includes sweet gum (*Liquidamber styraciflua*), hackberry (*Celtis laevigata*), red maple (*Acer rubrum*), several species of ash, and numerous other trees. At one extreme low hammock merges with marsh or cypress swamp, and at the other with mesophytic hammock.

MESOPHYTIC HAMMOCK. A typical and widespread type is the magnolia-holly-blue beech-ironwood forest with fairly rich moist soil, thick humus layer, and little undergrowth in the well-shaded lower levels. This is considered the climax growth for much of Florida from the lake region northward. South of the central portion of the peninsula the association is modified by gradual incursion of tropical trees; there is also extensive variation and shifting in dominants between the eastern and western extremes of the Panhandle. In regions of topographic and hydrographic change, mesophytic hammock merges suddenly or gradually with high pine, low hammock, and even flatwoods. In addition to the trees named above (*Magnolia grandiflora, Ilex opaca, Carpinus caroliniana, Ostrya virginiana*), greater or less admixture of the following is commonly seen: spruce pine *(Pinus glabra),* dogwood (*Cornus florida*), sugar maple (*Acer barbatum*), basswood (*Tilia caroliniana*), loblolly pine (*P. taeda*), mulberry (*Morus rubra*), etc. Pure mesophytic hammock supports a surprisingly meager fauna in view of the apparently favorable physical conditions that obtain.

UPLAND HAMMOCK. The topography is rolling or hilly, drainage is good, and the soil usually, but not always, calcareous. The lower strata of the forest are generally open except where succession toward more mesophytic conditions is taking place. The category embraces the following associations: high live oak or live oak–cabbage palmetto hammock, the red oak or red oak-hickory-sweet gum hammock of the northern part of the peninsula, and the various modifications of the red oak-beech-slippery elm forests of the red hills in the panhandle.

Though not highly characteristic, the fauna shows certain departures from those of the other hammock types.

TROPICAL HAMMOCK. A mesophytic forest of hardwoods, mostly West Indian in species, appears to be the climax association for the Florida Keys and the peninsula south of Palm Beach County on the east coast, Hendry County in the interior, and Lee County on the west coast. Hammocks of this type occur in the potholes or in old detritus-filled depressions in the limestone flatwoods, as insular elevations in the Everglades, along the banks of many creeks and rivers, and intermittently in the prairie land and buttonwood forests back of the mangrove swamps in the Cape Sable region and along the shores of Florida Bay. Fire, hurricanes, and cultural operations bring about extensive modifications, both in the successional sequences and in the character and distribution of the climax growth. The finest tropical hammock that I have seen is that covering most of the interior of Lignum Vitae Key, a small high island in Florida Bay off the lower end of Lower Matecumbe Key. Here many of the larger trees attain diameters of twenty and thirty inches; the soil is covered by a deep humus layer, and the dense shade prevents the development of the tangled undergrowth characteristic of many of the Key hammocks. The dominant trees are as follows: Jamaica dogwood (*Piscidia piscipula*), poisonwood (*Metopium toxiferum*), lignum vitae (*Guaiacum sanctum*), pigeon plum (*Coccoloba diversifolia*), mastic (*Mastichodendron foetidissimum*), and mahogany (*Swietenia mahogani*). In the freshwater succession the glade and cape hammocks merge gradually, through live oak–cabbage palmetto hammock or gumbolimbo-poisonwood savanna, with saw-grass marsh; in the saltwater succession, through buttonwood hammock or prairie land, with mangrove swamp.

Aquatic Situations

Ponds

FLUCTUATING PONDS. Shallow, muck-bottomed drainage basins may persist as ponds throughout an abnormally wet year, but they become almost or wholly dry during a normal dry season. True hydrophytic vegetation is absent or restricted to a few quick-growing or very hardy annuals. When the water disappears, the permanent fauna either burrows into the muck bottom or migrates overland to escape desiccation.

SINKHOLE PONDS. These small, usually circular depressions are produced by solution and collapse in underlying limestone. The depth ranges from ten to fifty feet—or more, on rare occasion; the average for the majority of sinkholes in the northern part of the peninsula is probably about fifteen feet. A marginal zone of rooted floating plants usually extends out to a depth of six or seven feet; emergent and submerged vegetation is scarce. The surfaces of some ponds become covered completely with duckweed (*Lemna spp.*), or with duckweed, mud-mary (*Bruneria*), and mud-midget (*Wolfiella*).

FLATWOODS PONDS. These ponds are local depressions in wire grass and palmetto flatwoods. There is usually a broad, more or less bare zone of fluctuation bordered by a narrow ring of semi-aquatic trees or by a dense saw-palmetto "ledge." The vegetation of the basin may be wholly composed of pond cypress (*Taxodium ascendens*), of cypress and black gum, or of a mixture of emergent herbs and grasses—maidencane (*Panicum hemitomon*), pickerelweed (*Pontederia cordata*), arrowhead (*Sagittaria lancifolia*), and various rushes (*Juncus spp.*).

CHARA PONDS. A peculiar variation of the flatwoods pond habitat is seen in the broad, shallow, irregularly shaped depressions of the west coast flatwoods. In many of these almost the only plant is a species of *Chara*, which grows in tremendous abundance, and despite radical changes in water level it often fills the entire pond basin. Though not a common type, chara ponds are worthy of mention because of the huge turtle populations they sometimes support.

HAMMOCK PONDS. A rather characteristic type of pond is found in hammocks and in cleared hammock land and is fed by springs, seepage areas, or runoff from the surrounding woodlands. The depth is seldom more than five or six feet, and emergent vegetation is frequently well developed, the commonest species being cattail (*Typha spp.*), water oleander (*Decadon verticillatus*), buttonbush (*Cephalanthus occidentalis*), willow (*Salix nigra*), and black gum. The water is usually permanent, and seasonal fluctuation is much less marked than in the flatwoods ponds. Floating and submerged hydrophytes are commonly found, especially duckweed, liverworts (*Riccia spp.*), bladderworts (*Utricularia spp.*), and coontail (*Ceratophyllum spp.*). Such ponds are common to upland, mesophytic, and low hammock, and in addition to the aquatic fauna that they support there is generally at their margins a concentration of the forms inhabiting the adjacent hammock.

Lakes

In the youthful drainage system of Florida, developed as it is upon a karst topography of relatively slight elevation, lakes are a conspicuous element. From the standpoint of geological origin, several types of Florida lakes may be distinguished. Solution lakes are common in the limestone regions of central and western Florida and are produced by the formation and confluence of a number of sinkholes. Such lakes often develop along the courses of old streambeds. Consequent lakes are probably not rare in central and southern Florida; the occurrence of several species of fish with marine affinities in certain isolated lakes in the interior lends support to the view that these lakes may be relict depressions. Oxbow lakes are present in the floodplains of most of the master streams. In the Everglades and in the mangrove swamps of the Cape Sable region there occur broad, shallow brackish or freshwater lakes of uncertain history; these may be partly consequent and partly of solution origin, or to some extent may have originated through differential land building.

Because of their larger size it is much more difficult to propose a satisfactory classification of lakes than it is for ponds. Some of the lakes appear to be composite in origin, and their present basins to have been formed by subsequent solution about an old consequent basin. Moreover, most of the larger lakes show all degrees of succession toward terrestrial conditions in some part of their basins. Thus the fauna of a given lake is very apt to include all the aquatic forms of the surrounding region.

Marshes and Swamps

FRESHWATER MARSHES. Marsh formation in Florida is rapid. The filling up of the lake basins is not contingent upon the accumulation of organic debris on the bottom alone but is accelerated to a great degree by growth of matted floating vegetation at the surface. Tangles of bonnet roots (*Nuphar luteum*) bound together by the long adventitious root systems of pennywort (*Hydrocotyle ranunculoides*), maidencane, and other hydrophytes form massive rafts or floating mats on which numerous aquatic and semi-aquatic herbs, shrubs, and even trees quickly become established. Saw grass (*Cladium jamaicensis*), whose buoyant root mass is capable of supporting the weight of the plant, is also effective in the building of floating islands. Several types of marshes may be distinguished—chiefly on the basis of the dominant plants. In many areas, notably in the Everglades and in the extensive marshes around the headwaters of the St. Johns River, the vegetation consists almost entirely of saw grass. "Prairies" of bonnets, water lilies (*Castilia odorata*), and water hyacinths (*Eichhornia crassipes*) are common, especially in central Florida. In the central lake region there are numerous marshes of maidencane, water oleander, pickerelweed, and arrowhead. The nature of the fauna of a given marsh is probably determined more by the degree of fluctuation in water level than by any other factor.

SALT MARSH. This association forms a coastal margin of varying width in most of the upper half of the state; it is most extensive along the estuaries of northeastern Florida. On the flats that lie between high- and low-water marks, the grasses *Spartina alternifolia* and *Distichlis spicata* are the dominant—and often only—plants. Where the sandy or muddy soil is inundated only by springtides or during storms, the aquatic grasses are replaced by rushes (*Juncus rhomerianus*), saltwort (*Batis maritima*), saltbush (*Baccharis spp.*), and the samphires (*Salicornia bigelovii and S. virginica*). Where salt marsh merges with flatwoods or live oak–cabbage palmetto hammock the transition is usually abrupt, but the lowering of salinity as the rivers and estuaries are ascended is made manifest by the gradual replacement of the typical salt-marsh grasses by *Spartina cynosuroides,* rushes, and finally by cattail, sawgrass and the characteristic freshwater plants. The salt marshes of the northern Florida coasts are represented in the southern part of the peninsula by mangrove swamps; the slight differences in the faunas of the two associations are probably due more to geographic than to ecologic factors.

FLUVIAL SWAMPS. The vegetation of a river swamp may be almost wholly composed of bald cypress (*Taxodium distichum*); but in northern Florida it is usually an association of several species dominated by cypress, tupelo gum (*Nyssa aquatica*), or ogeechee lime (*Nyssa ogeche*). The water level, though variable, is generally higher and more constant than in nonalluvial swamps, low hammocks, and marshes, and conditions that are essentially aquatic prevail during most of the year. The most abundant fauna occurs in those swamps in which fluctuations of water level are least radical.

BAYHEADS. Dense, tangled, thicketlike swamps in wet, nonalluvial soil or around the headwaters of the smaller creeks. The vegetation consists chiefly of small trees and shrubs. In the peninsula, sweet bay (*Magnolia virginiana*), black gum, wax myrtle (*Myrica cerifera*), red maple, dahoon holly (*Ilex cassine*), and loblolly bay (*Gordonia lasianthus*) are usually the dominant species. In the titi swamps of the Panhandle the following forms are common: titi (*Cliftonia monophylla* and *Cyrilla parvifolia*), he-huckleberry (*Cyrilla racemiflora*), wax myrtle (*Cerothamus inodorus*), Virginia willow (*Itea virginica*), and dog hobble (*Leucothoe axillaris*). The soil is usually acid, and sphagnum beds are often present.

Prolific growth of smilax, grape, and Virginia creeper, and the interlocking limbs of the crowded trees, exclude light to a great extent and render some of these swamps almost impenetrable.

MANGROVE SWAMPS. Much of coastal South Florida and the Keys is covered with mangrove swamp—a common association in the marine-littoral of most tropical and subtropical regions. The dominant tree is the red mangrove (*Rhizophora mangle*), which grows in dense stands from well below low-water mark to somewhat above reach of the highest tides. The transition from swamp to hammock or dune conditions is marked by admixture of black mangrove (*Avicennia nitida*), buttonwood (*Conocarpus erectus*), cocoa-plum (*Chrysobalanus icaco*), saltwort, sea-grape (*Coccoloba uvifera*), and saltbush (*Baccharis spp.*). Along many of the streams and at the edge of the Everglades the mangrove swamps appear to be retreating before the advance of saw-grass marsh. On the east coast the mangrove swamps terminate in northern St. Lucie County and on the south end of Merritt Island, Brevard County; on the Gulf coast they extend to the Suwannee River delta.

Streams

LARGER STREAMS. Nearly all the larger streams of Florida derive their water from sources within the political boundaries of the state—in part from peat-filled swamps, lakes and marshes, and in part from numerous large springs. The tributaries of the Apalachicola River penetrate the Piedmont of Georgia and Alabama, but this is the only Florida stream that is of any importance in the coastal plain drainage, and the only one that bears any great quantity of silt. The weak coffee color, characteristic of the water of most Florida streams, is due to the presence of organic acids collected in swamps and bayheads. Where the water is very acid, submerged and floating vegetation is usually not abundant, and the fauna is generally poor. Optimum conditions apparently exist in those rivers that run over ledges of exposed limestone or that receive most of their water from calcareous springs.

SPRING RUNS. Along the west coast these calcareous streams fed by one or more large springs constitute the major streams in the drainage systems of several extensive areas. The water is basic to circumneutral, depending upon the amount of swamp water introduced by tributaries. The bottom is rocky or sandy, and submerged vegetation (*Vallisneria, Philotria, Sagittaria, Ludwigia, Ludwigiantha*) often grows in broad, deep beds that are populated by snails, crayfish, and the larvae and nymphs of many insects.

SMALL STREAMS. The small streams of Florida are difficult to classify with consistency. The extremes of physiognomy are perhaps exhibited by the sluggish, detritus-filled, highly acid swamp streams and the clear swift brooks of certain of the Apalachicola Ravines, whose beds are often strewn with gravel and small boulders and which have much the facies of mountain trout streams. Flatwoods streams are typically shallow creeks or brooks with acid water and sand bottoms; where the flow of water is permanent, thickets and bayheads usually line the banks, but where seasonal fluctuation is marked, the vegetation may consist solely of semi-aquatic herbs and grasses. Hammock streams usually flow over beds of bare sand or gravel and have little aquatic vegetation beyond an occasional patch of *Ludwigia, Ludwigiantha,* or sphagnum. The pH varies from quite basic to circumneutral or slightly acid. Many of the time rills and rheocrene springs are bordered by bogs and seepage areas that support lush growths of lizard's-tail (*Saururus cernuus*), jack-in-the-pulpit (*Arisaema acuminatum*), begonia (*Begonia cucullata*), and the ferns *Lorinseria areolata* and *Dryopterus floridanum*. These rank beds of mesophytic herbs appear to be the optimum (and in my experience, the only) habitat of the gulf coast red salamander (*Pseudotriton montanus flavissimus*). The aquatic herpetofauna, however, offers little grounds for distinguishing among the several types of small streams.

SLOUGHS, CANALS, AND DRAINAGE DITCHES. These include some of the most fertile and prolific aquatic habitats in Florida. The water is acid to basic, and various species of submerged, floating, and emergent hydrophytes often grow in the greatest profusion. There is usually an abundance of fish, especially cyprinodonts, and invertebrates are numerous. Long reaches may be covered with the water hyacinth, and often the water is choked with coontail, *Cabomba, Myriophyllum,* bladderworts, *Potomogeton,* or *Proserpinaca.* The herpetofauna includes a number of fairly characteristic species that are sometimes individually very numerous.

Armadillo Dilemma

I have always thought that a reasonable amount of emotion is inevitable in zoology, provided the worker is involved with species other than white mice. Armadillos have been astir since northern Florida's early spring, ambling through the hammocks, endlessly tilling the forest floor. I write about them not because they upset flowerpots or are so popular at Florida barbecues or have lately become virtually the national animal of Texas, but because my feelings about them have become peculiarly schizophrenic.

There was a time when armadillos seemed to me among the most beguiling of mammals. I first met them when my father read Kipling's *Just So Stories* to me and we got to "The Beginning of the Armadillos," in which it is suggested that they arose when the tortoise and the hedgehog combined their special adaptations to foil Painted Jaguar. Nothing I later learned about armadillos has changed my view that as long as they stay in their natural range they are just great. They and their edentate relatives (toothless, or nearly toothless mammals) are the most otherworldly looking terrestrial mammals we live with, or have ever lived with, on earth.

In South America, armadillos range in size from the giant armadillo—five feet from nose to tail tip, with claws up to four inches long—down to the elfin pichiciego, or fairy armadillo—a five-inch molelike creature that carries around a flat scrap of shell like a little mat on its back. And look at the tree sloths: nothing could be more fey than they. The giant ground sloths of a few thousand years ago also had a storybook look, as did the ponderous glyptodonts, some of which had big spike-studded spheres at the ends of their tails. So it is not just *Dasypus novemcinctus*, the nine-banded armadillo, that looks weird. It is a whole orderful of animals with widely differing habits and all looking as if Dr. Seuss had designed them.

Back in the 1930s, when I studied mammalogy, there were no armadillos in northern Florida. Outside Florida there were none anywhere in

From *Animal Kingdom* 85, no. 5 (1982).

the United States except southernmost Texas, New Mexico, and Arizona. Today peninsular Florida armadillos are more numerous than any other mammal bigger than a squirrel.

The introductions that planted this explosive population occurred fifty or sixty years ago, when three groups of armadillos escaped in Dade County and on Cape Canaveral. The first was in 1920, when some armadillos got out of a Hialeah zoo. Another lot, it is said, escaped from a private zoo in Cocoa during a 1924 hurricane. Two years later and a few miles farther north, at Titusville, a circus truck overturned, and more of the creatures were freed.

The exact numbers of animals involved in these releases are not known. The total was certainly no more than thirty or forty. But it doesn't matter, really. Even if only one pregnant female had come in, her issue—if each lived out its span—could have overrun the state. Nine-banded armadillos bear four young per litter. Born of a single egg, they are identical, thus all the same sex. This is not bad, because any female is just as likely to produce one sex as the other, so the sex ratio works out to fifty-fifty. Assuming that every female breeds for three years before she dies and that there is plenty of habitat and food and no predation, the descendants of one pregnant colonist would, after forty years, number 436 billion. The reasons they aren't all here are that some predation does occur, particularly on the young; armadillos are killed by the thousands on highways; and in some localities they simply have eaten themselves out of house and home.

Meanwhile, however, they have spread throughout the peninsula, into West Florida and northward on the coastal plain as far as Macon, Georgia. During the same time they were extending their Florida beachheads, other armadillos were moving eastward from Texas along the Gulf coast, and now this salient has reached the Florida Panhandle. Whether the two sets of expatriates have met there, I don't know. In any case, the peninsular armadillos are my present subject.

I have been closely associated with the nine-banded armadillo in Honduras, Nicaragua, and Costa Rica, and nowhere have I seen it as numerous as in Florida. In the 1930s any dead creature you saw down the road was likely to be a possum, coon, skunk, fox, or house cat. Today, in summertime at least, it is far more likely to be an armadillo.

Why did the new Florida colony spread so intemperately? Put another way, if *Dasypus* finds Florida so agreeable today, why did it abandon the region a few thousand years ago after flourishing there during Pleistocene times? There were three lines of armadillos in Florida then: there were glyptodonts—two-hundred-pound armored creatures, fourteen feet from blunt nose to end of tail; a set of hulking butt-headed species belonging to the genus *Chlamytherium;* and other, smaller armadillos closely related to today's nine-banded visitor and placed by paleontologists in the genus *Dasypus.*

Having weathered the changing Pleistocene and survived until about six or seven thousand years ago, why did all the Florida armadillos disappear? The same can be asked of much of the other late-Pleistocene fauna—giant sloths, elephants, camels, dire wolves, saber-toothed tigers, and a host of others. They all left us, too, and the cause, though much debated, has never really been identified.

With *Dasypus* the mystery deepens: the remigrant armadillo clearly finds Florida a delightful place. It has spread and bred with such verve that you might think that ever since the Pleistocene extinction its kind has been down there in the Southwest yearning to get back.

As to why it quickly became so abundant upon reaching Florida again, the answer is probably the same one advanced to account for the spread of any introduced species. The immigrant simply finds itself free of traditional frictions—predators, parasites, competitors—and if it can stand the physical environment, it goes into an orgy of survival and dispersal. In this case, whatever killed off the post-Pleistocene armadillos is less lethal for the new immigrants.

Certainly armadillos must be finding predation less severe in Florida than back home. Predation is languishing in Florida nowadays. Few meat-eaters big enough to chase down and kill an armadillo remain.

Not much is known about who preys on the armadillo, but it is probably eaten by any carnivorous creature able to catch it and penetrate its armor. The armor is not developed in the young ones, and if they stray from their mother's burrow they are no doubt taken by raptors, snakes, and all meat-eating mammals. But after the armor hardens the animal is protected from the teeth and claws of small carnivores.

The natural Florida predators capable of taking the adults are panthers, bears, and bobcats. Of four panthers killed by cars on Florida roads during the past five years, two had eaten armadillos. Most of the bear stomachs that have been examined contained them. But both of these animals are so reduced in habitat and numbers that they are not a significant deterrent to the growth of the population. The bobcat is the only other carnivore that, in a few areas, may make recognizable inroads. Bobcats occasionally roam our farm. From the porch of his house, our nearest neighbor saw a big one kill and partly consume two armadillos in our field. But in most of the territory the armadillo has occupied, even bobcats have become too rare to be a controlling factor. The sad fact is, there just are not enough predators left to keep order in the ecosystem, and there are armadillos all over the place.

When *Dasypus* began to spread in Florida, the citrus growers complained that the armadillos' burrows were drying out the roots of orange trees. Gardeners, nurserymen, and golf course maintainers wrote to the newspaper or telephoned the Agricultural Experiment Station in Gainesville for help. A friend who lives on the edge of San Felasco hammock, which has become an inexhaustible reservoir of armadillos, gets up at 4:30 every warm morning and spends a couple of hours in ambush with a spotlight and a rifle on his porch, watching for the armadillos that spoil his plantings.

I have come to resent the presence of armadillos as much as any of these people have, but for a different reason. My quarrel with them is not their abuses of man's works. It is their destruction of the natural order of the peninsula's broad-leaved-forest communities.

The floor of a Florida hammock is the liveliest subsection of the community. It is not a single layer in the ecological organization of the forest, but a whole set of layers, from new leaf litter and leaf fragments of decreasing size on down through mold, organic debris, and organic soil to the underlying mineral sand. Quietly and inexorably, the Florida armadillos are disrupting this system. The ruin is not noticed by many people. Most tend to feel all right about a forest if the trees and squirrels are there.

In earlier years I spent a lot of time in the hammocks with colleagues from the University of Florida: at night, headlighting for wolf spiders, spadefoot toads, and the other creatures whose eyes shine back at you; in the daytime, turning over logs or scratching in the leaf litter after salamanders, shrews, lizards, and the small burrowing kinds of snakes that you rarely come across in the open.

A procedure we used in ecology classes gave me a baseline for comparing the populations of the leaf-litter animals of those pre-*Dasypus* days with those of today. This was sampling with a Berlese funnel—a three-foot metal cone with a sieve tray set into it several inches below the top. The funnel was supported in a frame that held its narrow end over a jar of alcohol. We raked up a measured amount of leaf mold and litter, stacked it on top of the sieve, and bent a gooseneck lamp over it. This dried it slowly, from the surface downward. As drying proceeded, the unfortunate small creatures in the sample would follow the descending zone of moisture, and when they got to the screen they would fall into the jar of preservative. It was hard on the animals, but it was a pretty good way to sample leaf litter.

Anyway, in pre-*Dasypus* days a Berlese sample might yield anywhere from twenty to one hundred or more little animals. There were insects and their larvae of a dozen orders; millipedes; centipedes; isopods (such as sow bugs); snails; mites; daddy longlegs; and a perfect welter of spiders, from big lycosid wolf spiders and tarantulas that lived in silk tubes at the bases of the hammock trees to tiny money spiders that sang silently to one another among the leaves.

If you do such sampling today in the same places we did then—even if the site has resisted human inroads and seems to the eye unchanged except maybe for the increased sizes of the trees—your Berlese sample will often contain nothing at all. It will never reveal the old diversity and abundance. Wherever armadillos have come into the hammocks you no longer find salamanders or ring-necked snakes under the logs they should be under. You shine the eyes of maybe a couple of spadefoot toads where dozens ought to be. It is the same with crowned snakes, red-bellied snakes, and coral snakes, and even more conspicuous with skinks—both the big *Eumeces* (the genus to which most of our skinks belong) and the tiny ground skink, *Scincella lateralis,* once the most ubiquitous and abundant lizard in the woods of peninsular Florida.

Armadillos have brought about this appalling change in two ways. One is simply that the animals eat every living thing they come across in their indefatigable rooting. Armadillos are usually thought of as insectivorous. Actually, they are total omnivores, limited in the scope of their diet only by feeble dentition and a relatively small mouth. The stomachs of two road-killed Florida armadillos contained worm lizards six inches long; that of one from our farm was packed with mushroom fragments. They eat literally anything they can overpower and get down their gullets.

Their other adverse effect on the ecosystem is much more serious. Their assiduous rooting is destroying the organization and productivity of the leaf-mold stratum of the forest. In armadillo woods hardly a square foot of ground remains unprospected for long. Aeration of the leaf mold dries it out; prevents stratification and decomposition; makes it uninhabitable for leaf-mold biota, from microbes to vertebrates; and wrecks it as a conversion zone for forest detritus and nutrients.

I don't know what the long-range effect of the disruption of the forests will be. Perhaps the armadillos will root themselves out of profitable feeding ground before irreversible changes occur. There is evidence that this has happened in some North Florida hammocks. Meanwhile, to those of us who spent much of our youth scratching in the hammock leaf mold, the woods of central Florida seem an almost empty framework for a fauna almost lost.

There are useful morals in the melancholy mischief of the Florida armadillos. One is that animals and plants ought not be moved around to places outside their natural range. Another is that trapping bobcats in Florida ought to stop. A third is that the natural world can be spoiled in very subtle ways.

*W*ater *H*yacinths

Singly, the inflorescence of the water hyacinth is a thing of beauty, a spathe of incredibly fragile flowers of an ethereal lavender. But the full impact of hyacinth bloom comes when you see it spread over ten or a hundred acres, or down both sides of a long reach of a river—or in the canals beside the roads that cross Paynes Prairie. Two highways cross the prairie between pairs of broad ditches, and these used to stay completely covered with water hyacinths. In April and May you drove the two miles over the savanna between lavender ribbons. I have seen tourists, evidently on a first springtime trip to Florida, stop with a screech of tires as they came to the northern edge of the Prairie and stare agape at the two strips of color converging ahead of them. Sometimes a woman would get out of the car, run down to the bank, and hastily pick an armful of flowers to take away. That is a melancholy thing to see, because the petals of hyacinths are so filmy that they quickly melt into blobs of pulp when the flowers are picked.

To most people in Florida and other warm parts of the United States, the water hyacinth is just a grievous pest whose sins are only fleetingly compensated for by the spectacular show of bloom that the plants put on each summer. And certainly *Eichhornia crassipes* is a pest—one that has spread and grown disastrously as man progressively overfertilized the waters of the world with his wastes. But the natural history of pests can be just as engrossing as that of more amiable organisms, and often more so. The water hyacinth is more than a mere blight on southern waters. It is a plant of surpassing ecological interest.

I first became involved with water hyacinths when I was an undergraduate biology student at the University of Florida and aspiring to take up graduate work on the southeastern reptiles and amphibians. The hyacinths drew my interest not because they had pretty flowers or because they were a threat to freshwater navigation, but because the floating rafts they

Adapted from *Animal Kingdom* 87, no. 5 (1984); and *Animal Kingdom* 88, no. 1 (1985).

make were a productive collecting ground for reptiles and amphibians not easily found elsewhere.

If you keep your eye on a herpetologist for a half-hour or so, you are likely to see him reach down and turn over a rock or log or a sheet of old roofing metal. Herpetologists turn objects over because many kinds of small cold-blooded vertebrate animals—snakes and lizards, most salamanders, many kinds of frogs, and, down in the tropics, practically all cecoelians—are more readily found that way than by any other method. Back in the days when the University of Florida was a hotbed of whole-animal biologists, we worked out a collecting technique that was even more productive than turning over logs. It was rolling up hyacinths. I would like to know how many ergs of effort my fellow graduate students and I expended, altogether, dragging hyacinths out of the waters of the county. Not only was the technique productive, it required no apparatus at all—merely strong motivation, physical stamina, and a high tolerance to muck-itch and to hot-bugs.

The hot-bugs were pretty bad. They drove some tentative naturalists of the time into other academic lines. The hot-bug is a hemipteran of the genus *Pelocorus,* a fat, attractive little insect handsomely marked with green and brown. Hot-bugs are abundant members of the water hyacinth community of Alachua County, Florida, and they have materially held up study of the water hyacinth ecosystem. The trouble with a hot-bug is that it intemperately plunges its proboscis into anything that intrudes upon its privacy. Actually, the bite is no worse than a bee sting, but it seems much worse. Bees sting openly, out where one is able to see what has attacked. When a hot-bug bites, you may be navel-deep in hyacinths and muck, where it is neither possible to see what bit or to get quickly out of the place. Some otherwise strongly motivated zoologists take years to reach a professional level at which they can undergo a hot-bug bite without panic. Some are never able to do so.

The other hazard of hyacinth-rolling is muck-itch. Exactly what this is I never have known. I only know that sometimes muck burns on contact and sometimes simply leaves one with a lingering rash. In either case it doesn't hold a candle to hot-bugs for keeping timid people out of the water hyacinths.

But really, both together were a small disadvantage to balance against the harvest to be had from rolling hyacinths. And besides, a good population of hot-bugs in a hyacinth raft is a propitious sign, because hot-bugs are predators and are there for the purpose of preying on the host of smaller creatures that inhabit the hanging, fibrous root masses. The hot-bugs, in turn, are preyed on by bigger beasts, and so on, in a complex web of feeding relations that under good conditions may involve scores of species and thousands of individual animals in any single square yard of well-grown hyacinth raft. In South Florida waters James O'Hara found that a square meter of pond had up to eighty-four thousand invertebrate animals where hyacinths were present.[1]

To be most productive, hyacinths should extend only a short distance out from the shore, leaving most of the surface of the water open to sunlight so that the plankton can make food and produce oxygen. What we used to look for was a pond, lake, or ditch with a narrow fringe of mature hyacinths along its shore and with bottom contours such that a collector could wade out through the floating plants for ten or fifteen feet before he reached water more than waist deep. When we located such a place, two or three of us would line up facing the shore, and then all together would start lifting armfuls of hyacinths and throwing them on top of the mat, while at the same time walking slowly toward the bank. If the hyacinths were well-grown old ones, the long roots would tangle with stems and leaves and bind together a mass that grew like a rolled snowball. By the time this reached the shore it would be a ponderous, squashy but coherent bolster. We had to trundle it several feet farther up the bank, because otherwise the more agile creatures in it would slither and scramble to freedom with the water that rushed back down the slope. Once the mass was safely lodged on solid ground, we pulled it apart, plant by plant, and shook each hyacinth separately over a pan or a hard clean patch of ground to dislodge the clinging denizens.

As I look back on my junior year in college, when I changed from a major in English to one in zoology, one of the factors that I now see influenced my decision to change was the hyacinth fauna. It was being able to go out and predictably catch a whole lot of self-effacing little animals that most people don't even know exist. For all the decades since my undergraduate days, zoology classes at the University of Florida have been taken on hyacinth-rolling field trips. Besides being a good way to show students animals that they might not otherwise ever see, rolling hyacinths is also a good way to illustrate ecological relationships. Besides being individually diverse and numerous, the water hyacinth fauna furnishes an almost unique example—certainly the best in Florida—of an ecological organization in which a group of animal species is consistently associated with a single species of plant. Clear examples of this are hard to demonstrate. In Florida, however, if you go to the right pond or canal and industriously roll up, rake in, or dip out water hyacinths, you can usually count on catching an exciting array of reptiles, amphibians, fishes, and invertebrate animals, many of which are hard to find elsewhere and which nowhere can be found in such predictable associations.

For example, there are at least six species of snakes and a couple of kinds of salamanders that one could reasonably count on only in the hyacinth community. All of them live in other places, but just try to find them on any given afternoon anywhere other than in a water hyacinth raft. Besides these habitual denizens, there is a long list of other cold-blooded vertebrates, crustaceans, and aquatic insects that turn up regularly and sometimes abundantly in the hyacinth community. Ecologically, the association includes leaf eaters, detritus eaters, parasites, and predators of all sizes, and to some extent these subdivide the living space within the hyacinth raft, sorting themselves into strata, from the glossy leaf surfaces where anoles stalk blind mosquitos and green tree frogs yap, clear down to the nethermost hanging root tips grazed by gator fleas and little snails. For this extraordinary fraternity of creatures, vertebrate and invertebrate, air-breathing and aquatic, the hyacinth with its fibrous root mass, broad leaves, and cluster of upright petioles is hiding place, pasture, and hunting ground. The occupants obviously find in hyacinth rafts emoluments that enhance their survival and chances for reproduction, and so are bound together in a clear-cut ecologic organization.

The hyacinth association seems so natural and well established, in fact, that one finds it hard to believe that the host plant arrived in the southeastern United States less than a century ago. And no doubt the fraternity did exist all the time, at least loosely, in such other floating, raft-forming plants as pennywort and water lettuce, and in the masses of debris and vegetation that form floating islands.

At first glance, water lettuce seems the most likely native framework for the community now found among hyacinths. It, too, is a floating plant, and it is roughly similar to the hyacinth in size and shape. And yet you rarely find many animals in water lettuce. This may be in part because, in Florida at least, it grows mainly in streams and spring runs while hyacinths will grow in nearly any kind of water. Hyacinth is especially happy in the shallow ponds, lakes, marsh pools, and ditches in which the creatures prone to secrete themselves in floating vegetation are most abundant.

Before hyacinths spread so widely and crowded it out, however, water lettuce was more widespread and may have been more important as the natural asylum for the animals we now think of as the hyacinth fauna. In any case, as a many-leveled living space, hideaway, and hunting ground for small aquatic and palustrine creatures, hyacinths and water lettuce seem similar. Both form continuous, floating rafts; long, fibrous rootlets hang in dense masses from the bases of both; and, seen from above, the green decks that they make appear to be equally dense. If you put on a face mask and dive down beneath them, however, you notice an important ecologic difference between the two. Under a stand of mature, healthy hyacinths there is darkness. The only light there leaks in through scattered slashes where some recent accident has rent the deck or upset a plant. A water lettuce raft, on the other hand, is translucent, and a dim green light prevails beneath it. The lettuce leaves make less shade, and the yearly fall of debris never builds up the accumulation of organic mud that sometimes rests around the bases of old hyacinths, like a sort of false bottom up at the top of the pond. That this difference in the light that they let in is a fundamental one is indicated by the fact that the roots of water lettuce are green with chlorophyll while those of hyacinths are sooty black or dark purple in color. So, despite the similarity between hyacinths and water lettuce, there are important differences in the kinds of habitat they make; and that of the exotic hyacinth supports by far the more complex ecological organization.

Another place in which the hyacinth animals may have come together in pre-hyacinthine days was in floating islands. Floating islands used to be a regular feature in the solution lakes of northern Florida and an important factor in the ecological succession of lakes toward marshes. The usual fate of a lake or a pond is to obliterate itself by depositing debris that gradually fills the basin from the bottom up and finally converts the site to a swamp or a marsh. In some Florida lakes this process is accelerated by the formation of floating rafts of muck and vegetation.

The floating islands of Florida are substantial enduring parcels of land that cruise before strong winds and support a fauna and flora that travel with them. They seem a wholly natural part of the landscape. The details of their origin have never been completely figured out, but an important part of their framework is, in some cases at least, contributed by spatterdock—or bonnets, as *Nymphaea* is known in Florida.

Bonnets have enormous roots, and when the plants die the roots rise to the surface, bringing up masses of peat. These rafts apparently are buoyed by decay gases that accumulate in and under them. As these rafts consolidate and soil forms on them, grasses, sedges, and herbaceous plants take root there, and eventually maple, saltbush, and willow saplings appear. These may eventually grow into little copses with herons and grackles nesting in their limbs. In times of low water these islands may rest on the bottom, but when the water level rises they often break free again and cruise back and forth across the lake. Orange Lake in Marion County, Florida, is famous for its floating islands. I once watched an island rookery break loose near McIntosh and on a strong west wind cruise out of sight down the middle of the lake toward Cross Creek ten miles away. As you move around the edges of Orange Lake, you may see patches of bubbles, some the size of your boat, others many times as big. The catfishermen say these mark the positions of incipient floating islands destined to rise to the surface.

My first encounter with a floating island was back in my undergraduate days, when one sailed away with me walking around on it. I had shoved my little duck boat onto an edge of what I took to be a peninsula jutting from the shore. I had stopped there because I saw a litter of turtle remains near the edge of the water. When I got out to inspect the bones, I noticed a quaking feel to the ground; but it wasn't alarming, and I walked over to the turtle debris and examined it.

The shells of six turtles were there, all softshells and all represented by only the carapace and the plastron, joined by the bridge; the contents of each had been neatly removed. I figured that either otters or raccoons must have left them. Since then I have seen quite a few such piles but never have learned which of the two suspects is littering Florida islets with turtle carcasses. Whichever it is, it does the same thing with hard-shelled cooters, the three local species of *Pseudemys*. Just how the poor turtles are extracted from their shells is an added mystery. In the case of the softshell remains, my snap judgment was that an otter had to have been responsible, because I couldn't—and still can't—see how a coon could catch a softshell, whereas an otter can catch just about anything.

Anyway, as I pondered the puzzle it was slowly borne into my mind that a gap was appearing across the base of the peninsula. The weather was squally, and a sudden flurry of cold wind was bending the willow and wax myrtle trees on the little patch of land. Under my dumbfounded gaze the supposed peninsula became an island and set out before the wind on a course across the lake.

I was bemused by this seeming violation of natural law but not seriously alarmed, because my little boat was there to take me back to shore . . . or so I thought. But as I looked toward the boat for comfort, I saw the ground moving out from under the bow, and before I could get to it there was six feet of cold water between us. It was duck weather that day, too cold for swimming, which, short of bellowing for help from random sources, seemed my only recourse. But before I got around to yelling, another swirl of the northeaster pushed the boat back against the island. I moved cautiously to its quaking edge, got into the boat, and resumed my trip around the lake shore.

Within the bodies of floating islands you can find most of the hyacinth animals. With several fellow students I once spent an afternoon running screen-bottom boxes under little floating islands, and under the edges of big ones, and then pulling the mass apart while holding the screen beneath them. With hard work we eventually got both *Pseudobranchus* and *Siren*—the two eel-like salamanders that live in hyacinths—and a couple of *Farancias*—the aquatic mud snakes that like to eat those eels.

We also found small numbers of most of the other members of the hyacinth community: the red-bellied mud snake, some little soft-shelled and stinkjim turtles, and the common kinds of crustaceans, snails, and aquatic insects. But all of these were far less abundant and less intimately grouped than they would have been in an equal volume of good water hyacinth raft.

It is hard for me to be objective about hyacinths. Besides the interest they have always held for me, I have suffered personally at their hands more than once. They blew in on my duck blind one freezing evening, with a sudden change of wind; and I spent five hours, the last four of them in the howling dark, clawing my way out to open water. One year down at Tortuguero, Costa Rica, the natural controls of hyacinths—whatever they are—relaxed, and the plants blocked eight miles of the waterway out to civilization. They stayed there for seven years and then, suddenly, the jam broke up, the hyacinths died and sank or drifted out to sea, and dugouts could travel the lagoon again. I have lost fish traps under shifting hyacinths, and they have broken trotlines I have set. Twice I saw my wild pointer dog, reckless in his vice of chasing swamp rabbits, dash out from the solid shore onto floating hyacinths, disappear from view, and come up thrashing miserably, his nose barely out in the air. Each time I had to wade out chin deep to bring him in. On another occasion I saw a good friend of mine disappear for a time under hyacinths. He was just down from Pittsburgh, not used to water hyacinths, and in a hurry to pick up some ring-neck ducks we had shot from ambush on the shore of a lake. The air temperature stood at nineteen degrees Fahrenheit that day, and the water was not much warmer. With hip boots on, and held up by pure enthusiasm, the friend ran quite a way out across the hyacinth fringe before he sank out of sight.

But the greatest grief that water hyacinths ever brought me was the loss of Lake Alice and Bivens Arm. Not that either place belonged to me in any material sense; and not that either is wholly lost. But both used to seem to me to rank among the wonders of the world, and much of their old magic is gone. Water hyacinths, in the virulent phase they get into when they grow near civilized man, were the chief cause of the sad change.

Bivens Arm and Lake Alice are both marshy solution lakes occupying shallow basins formed by collapse of the honeycomb limestone that holds up much of north-central Florida. The arm is a narrow-necked bay of the great plain called Paynes Prairie. Much of its shore is University of

Florida farmland; the rest is lined with private residences. Lake Alice is located wholly on the campus of the university. Before the hyacinths came, both places were superb examples of an exciting biological landscape: the Florida prairie-lake community.

The enchantment of these places is not their scope or grandeur but rather the intimate intensity of the diverse, noisy, passionate life that goes on in them. A good time to visit them is a morning in May, a golden-blue one after a short, hard shower of rain, when the slanted sunshine is still not hot enough to drive the snakes and cooters from their basking places about the edges of the floating islands and when snakebirds perch on snags with their wings outstretched to dry and alligators bellow back at passing jets.

Mornings like that stir the vocal animals of the marsh to celebrate with song. Gallinules keen and crackle then, least bitterns coo, and the shining male boat-tailed grackles creak, churr, and brandish their slim virility. An osprey, hovering high, peers down, hawkeyed, after fish in the shallows, then sails off, piping pitifully over a distant eagle he has seen. In the sunshine of such mornings red-winged blackbirds seem driven to drown with their bubbling melody the song of the mockingbird on a high sweet gum twig, the shouting wrens back in the woods, and the liquid *cheer-cheering* of redbirds in the pond-side fence rows. Frogs sing in daytime on May mornings after rain: cricket frogs *ik-ik* in teeming chorus; the sulfur-belly bullfrogs drum at one another across acres of pickerelweed; green tree frogs yap in ragged choruses among the buttonbushes; and common bullfrogs boom from the pools back under the willow and black gum trees. A rutting alligator slaps the water in a bonnet patch, causing a wood duck to hurtle off squealing through the hammock trees and a great blue heron to pick up and waft himself away, squawking in raucous indignation at the noise. A sora rail and a dozen gallinules all shriek together in simpleminded sympathy. A snakebird lurches off his drying snag and flaps away, heavily, like an *Archeopteryx* just up out of the lithographic limestone, *buzz-buzz-buzzing* as he goes; then suddenly he chops off his song by diving headlong beneath the surface of the pond.

Those are the overt, noisy workings of the place. Down in the tan water, algae soundlessly make sugar by the ton, and teeming fry of the year pick specklike plankton out of the water around them. A ten-pound bass in a patch of maidencane strains silently over a half-grown marsh rabbit that was smaller than his mouth but bigger than the stretch of his gullet. Beneath a field of lily pads an otter happily chases a hapless crappie, and out in open water the snakebird from the snag now follows with unexpected grace the frantic turns of a fleeing bullhead. Truculent mud fish tend swarms of finger-long young; warmouth bass fan fry in their nests in the beds of cress and drive off prowling water snakes, bream, and bungling congo eels. In the edge of the hammock the big softshell turtles come out and steal about among the trees, seeking to lay their eggs unseen by the keen-eyed fish crows, and a female alligator, watched only by the squirrels, ponders whether to nest on last season's mound or to raise another heap of debris for this year's eggs.

And those are May days, mind you, when the wintering host of coots is gone and when widgeons, teal, shovelers, and ring-neck ducks no longer dabble in the edges of the duckweed decks or graze the cabomba pastures. And in other years, when the willows, maples, and buttonbush trees have grown back after previous overuse by nesting birds, the trees again get frosted white with ibis, and there are five kinds of herons there chuckling and croaking about their nesting. In the rookery years the lake and the arm are fantastic places, flooded with movement, sound, and color. But even after the ducks are gone and in the off-years for the rookeries, when only bitterns, snakebirds, and little green herons nest separately about the floating islands, and when a couple of blue-winged teal left flightless after the hunting season peep together among the moorhens—even in those average times, the lake and the arm are wondrous places, or have been until lately.

The damage they suffered came from the interaction of hyacinths and man. Bivens Arm received insupportable injections of overflow from the city sewage plant and of nutrients from the University of Florida farm and animal pens; Lake Alice got farm runoff and a heavy load of phosphorus and nitrogen from the university sewage disposal system. Both responded with repeated blooms of plankton or submerged plants. These depleted the oxygen in the water, and fish died and floated belly-up in tens of thousands. Then the hyacinths began to grow, and before long the arm was so nearly covered with them that during days of offshore wind the people who live along the southern shore were unable to get their boats out to open water. Lake Alice, being smaller, less open to the winds, and even more heavily overfertilized, was wholly overgrown with hyacinths. As the months passed, life waned in the dark water under the opaque roofs of floating plants, and the once complex pond communities grew lean and sickly. From time to time people came with herbicides and sprayed the hyacinths. They died and sank, and their debris on the bottom made more pollution of a different kind and this dragged the biota further down.

In Bivens Arm a campaign of periodic poisoning finally brought the hyacinths under control, but not until the ecology of the place had been badly distorted. The University of Florida tried to poison the Lake Alice hyacinths, too; but the enthusiasm of their growth in the excessively fertilized water took the heart out of the project, and for years the lake lay there unseen under a magnificent field of water hyacinths said by some to be the most stupendously thrifty crop the College of Agriculture had ever grown.

Neither of these abused places is irretrievably lost. If they could be kept free of waste nitrogen and phosphorus both would regain at least some of their old ecologic diversity. The hyacinths of Bivens Arm are now under chemical control, and though the water periodically turns into pea soup, it nevertheless supports a populous though curiously simplified fauna dominated by snakebirds, bullheads, and badly oversized alligators. And as I write, the University of Florida has finally undertaken to get rid of the Lake Alice hyacinths by mechanical means. Airboats push the floating rafts within reach of a dragline with a specially designed clamshell that picks up the plants and drops them into dump trucks. Volunteers from conservation groups then come and lift out by hand the little clusters of hyacinths in the backwaters and beneath the overhanging trees along the shore.

But obviously, removing water hyacinths from these places is only treating a symptom of the pandemic malady of overenrichment; the freakish overgrowth of hyacinths in the county is merely a dismal result of growth without foresight. If the lake and the arm could be freed of excess nutrients they would soon cure themselves. But as long as the exudations of humanity pour in the hyacinths will riot and grow marvelously tall and crowd together joyously in hydroponic splendor. And each spring they will celebrate the spread of man with fields of lovely flowers.[2]

Triple-Clutchers

It has long been known that the Peninsula cooter, *Pseudemys* (*Chrysemys* to some) *floridana peninsularis*—a big dome-shelled turtle of peninsular Florida—is unorthodox in its nesting behavior. Like other turtles, the female selects a pleasant spot on the ground for her nest, wets it down with bladder water, scratches around awhile, then digs her egg cavity. Also like others, she digs with her hind feet, keeping the forefeet firmly planted. While digging she shifts her rear end back and forth to bring each hind foot alternately over the growing nest, which is slowly formed into an urn-shaped receptacle of a size appropriate to hold her egg complement.

The unorthodoxy of *peninsularis* is this: having once made a little progress in digging her main nest cavity, she interrupts that work from time to time and, swinging her body from one side of the nest to the other, scoops out two smaller, shallower side pockets. Ross Allen first told me of this strange routine. He had witnessed it in his turtle pen at the Reptile Institute in Silver Springs, Florida. I later described it in my *Handbook of Turtles*.[1] Since then I have watched the process many times.

There is nothing aimless or accidental about the making of the side pockets. Using first one back foot and then the other, the turtle works at the central site, shifts her hind quarters over a side hole and makes a few scoops there, then returns to the central hole. After a long period of this scooping and shifting she stops, lowers her tail into the central cavity, drops a few eggs, moves to one of the side holes and leaves an egg, then returns to the main nest. By the time all the eggs have been laid, there will be maybe eighteen in the central nest and one, two, or three in each of the lateral pockets.

All the other turtles of the world, whether they lay two eggs or two hundred, dig a single cavity to house them. Whatever could have gotten into this one subspecies, stuck way down on the Florida peninsula, to evoke

From *Animal Kingdom* 86, no. 2 (1983).

this inexplicable but very purposeful-looking variation in a ritual of the ultraconservative group to which she belongs?

Since I learned of the phenomenon I have been uneasy over my inability to see any adaptive value in it and even more so over the rest of mankind going blithely along, as totally in the dark as I am but unconcerned about the mystery. Partly, I judge, this strange indifference comes from the vague feeling people seem to have that the quirk is some sort of nervous disorder—one that causes the turtle to fumble her eggs and sloppily spill a few into a couple of casual pits that her hind feet have absentmindedly dug. But that is just escapism.

The few people who have responsibly pondered the matter have embraced the decoy-egg hypothesis. For years I leaned to this idea myself. According to that notion, when an egg predator divines a turtle nest it snuffles and roots around after the eggs, locates those in the side pockets, eats them, and goes away—either filled up or under the impression that it has eaten an entire cache of eggs. That appears to be the ranking hypothesis.

A defect in that idea is that one of the main despoilers of cooter nests, in Florida as elsewhere, is the raccoon. Anyone familiar with the shrewdness, industry, and insatiability with which that animal forages and feeds knows that two or three eggs in little side pockets would neither deceive nor satiate a coon. Nevertheless, the decoy theory has remained the ranking one, and even Peter Pritchard, in his monumental *Encyclopedia of Turtles*, says—rather despondently, I am happy to see—that there are those who believe the adaptive value of the side nests to be predator distraction.[2]

I used to puzzle over this to the point of being considered simple-minded by colleagues. Then, during the 1970s, some things happened that confused me even further. One was my slow realization that, in Florida, fish crows may be even more important egg predators than raccoons are. Crows can frequently be seen flying around carrying things in their beaks. Usually the objects are unidentifiable.

A few years ago, however, out at our farm we noticed that the things in the beaks of the crows flying across the pond were often white and about the size of the last joint of one's thumb. The idea that these were turtle eggs gradually sank in. Sure enough, when the peak times (May and November) of the split nesting season of *Pseudemys* came along, the numbers of crows carrying white objects across the pond always rose noticeably.

At first this led me to look more leniently on the decoy theory. Maybe the side pockets evolved as a way to distract or satisfy the hunger of fish crows—a lesser hunger, surely, than that of a coon. The flaw in this notion was that crows probably couldn't dig very well. Why waste two or three eggs on a predator that can't dig down to the main cache anyway? But that idea was dispelled when I saw a crow find, dig out, and carry away, one by one, the whole egg complement of a big soft-shelled turtle that had nested just outside my study window.

The crow's digging technique was a peck-and-side-flip movement. It made no effort to open up a big access hole, but restricted digging to the neck of the cavity, which was only a couple of inches across. The top eggs were about three inches beneath the surface; when the crow reached them it tweaked one out, lodged the egg in its beak, and flew away with it.

The eggs of a soft-shell are spherical, not oval like those of *Pseudemys;* the shells are brittle, like thin china. They are a bigger beakful for a crow than those of hard-shell turtles. Nevertheless, the crow (or crows— there could have been several taking turns) carried them all away and did something with them out in the forest.

So much for the fish crow as the predator in the decoy theory. The nest site of the Peninsula cooter is usually more conspicuous than is the single scratched-over nesting place of a soft-shell. Thus the peck-and-flick technique of a crow should be even more effective in digging out the shallower nest of the hard-shell cooters. And as for the crow's hunger being satisfied by "decoy" eggs in the side pockets, that is nonsense.

In recent years, I have seen dozens of three-hole nests that had been opened and robbed by crows. They are easy to identify. Unlike coons, crows don't excavate the whole premises but remove only the filler sand from the nest, leaving the three holes exactly as fashioned by the cooter. And crows don't leave a pile of empty shells, as coons do, because crows carry away the eggs to eat or hide. Besides this circumstantial evidence, every turtle season we watch the resident crows of our farm flying, eggs in beak, steadily back and forth across the pond. They travel between the wooded shore and a little island where old alligator nest mounds are favorite nesting places for turtles.

Thus the three-nest aberration is still a puzzle; and to further complicate the mystery, it has been learned that two other turtles of the genus *Pseudemys* share this eccentricity. One is the Florida cooter, *P. floridana floridana,* a resident of the southeastern coastal plain and close relative of the Peninsula cooter but different from it in color and markings. The two merge and interbreed in northern Florida, but if you mixed a dozen from Orlando with a dozen from Valdosta, Georgia, anybody could sort them out.

The Florida cooter is abundant in the Okefenokee Swamp. There one May day when herpetological colleagues Peter and Anne Meylan and I were eating hamburgers near a window in a little restaurant at Stephen C. Foster State Park, a big brown Florida cooter came up out of a ditch and in plain view dug a three-hole nest just like the one the Peninsula cooter digs. She put one egg in one of the side holes, two in the other, and a lot more in the central hole. Then she covered all three holes and went away. Since then I have been back up to the swamp several times during the nesting season. I have found dozens of three-hole nests and watched three other turtles dig them. There is no doubt that the custom is ingrained in the southern Georgia subspecies.

Then there was an even more sensational development. R. C. Smith, a zoology student at the University of Florida, found Suwannee cooters nesting near Dunnellon; they, too, dug triple cavities. This was doubly stirring because nobody had ever been able to find Suwannee cooter nests. When R. C. found them it was the climax of a long quest by zoologists from the university and the Florida State Museum.

One able graduate student of mine, Dr. Crawford Jackson, studied the ecology of *P. concinna suwanniensis* for his doctoral dissertation. Despite diligent searching—in the end it grew almost frenzied—the only eggs Crawford ever got came from the oviducts of dead females. So the reason none of us had never seen the Suwannee turtle dig three nest holes was that we had never seen one dig any holes. This was probably partly because the Suwannee River turtle nests in March and April—earlier than the others—and its emergences are usually restricted to days following heavy rain. Another distracter could have been its tendency to climb steep banks when emerging to nest. Other Florida turtles choose more gradual approaches, and we used to confine our searching to woods and fields along such shores when we looked for nests of *suwanniensis*.

Because the Suwannee turtle strongly differs from the *floridana* group —in conformation, markings, and behavior—its turning out to be a three-hole nester must surely tell us something about the ecological background of the habit. What we have, now, is a trio of southeastern cooters—two of them closely related geographic subspecies, the other a very different creature restricted to Gulf-river habitats and assigned to another species—violating the ancient and universal chelonian reproductive norm by distributing their eggs in three holes, to no understandable adaptive end.

We are thus back where we started—or maybe even worse off—in the search for an explanation. I can't see any glimmer of an ecological advantage in the iconoclastic practice. Yet I am not willing to shrug it off as idle irresponsibility, as three kinds of turtles millions of years apart in evolutionary origin violating rules just for the fun of digging a couple of extra holes in the ground.

I can only suggest this possibility: maybe, in today's post-Pleistocene ecosystem, the trait is devoid of adaptive value. Maybe it is just a behavioral relic that reflects Pleistocene habitats in which these three cooters were plagued by a different set of egg predators among creatures known to us only as fossils or not at all. Going down the list of possible Pleistocene candidates for the role, I can think of none so small of appetite or dim of wit that two little side nests would distract it from the mother lode of eggs. That is paleoecology. I am not a paleoecologist. No doubt the people versed in the lore of that field lose little sleep over three kinds of turtles breaking the behavioral regulations of their kind, but I wish they would turn their attention to it.

The Gulf-Island Cottonmouths

For several years and from many quarters I heard vague tales of appalling numbers of snakes inhabiting the little Gulf islands in the vicinity of Suwannee Sound. Visiting sportsmen returned with incredible stories of their numbers. The inhabitants of the little coastal towns of the neighborhood, though at great variance in their interpretations of the taxonomic status of the form, all agreed that the island brand of snake possesses a biotic potential more vigorous, a venom more lethal, and a disposition more treacherous and vindictive than any other North American reptile.

An attempt to formulate a coherent concept of the serpent or serpents responsible for the harrowing reports met with little success. The more conservative of narrators identified the species as copperhead; the more imaginative pronounced it sea cobra. Between these extremes of nomenclature were proposed such picturesque names as stump moccasin, stump-tail viper, saltwater rattler, and mangrove rattler. I discussed the matter at some length with an old fisherman who had lived many years on one of the islands. His observations had led him to conclude that there were four kinds of snakes on the keys off the Suwannee delta—all equally deadly. The rarest of these he described as a rough-scaled tan snake with long stripes; for this creature he knew no name. Then there was the common black stump moccasin that used to eat his young chickens; the copperhead with bright colors and foul temperament; and, worst of all, the little green-tailed water rattler, which never attained a length of more than eighteen inches and from whose bite recovery was impossible.

Those of you acquainted with snakes will, perhaps, wonder why I did not immediately identify the species described in the fisherman's account. The first in his list could be none other than Clark's water snake (*Natrix clarkii*), and the last three are obviously stages in the pattern development of the cottonmouth moccasin (*Ancistrodon piscivoris*). In defense I can

From *Proceedings of the Florida Academy of Sciences*, vol. 1 (1936).

only remind you of the attitude of slightly pained though conciliatory unresponsiveness with which the professional zoologist always receives the reports of the amateur. He expects to learn nothing of importance and, consequently, rarely does.

But cottonmouths they were, and the establishing of the fact was an experience fraught with excitement as well as ecological interest.

On April 4, 1934, a group from the department of biology at the University of Florida embarked on a general collecting trip to the islands off Cedar Key. The party was composed of Buck Bellamy, H. K. Wallace, John D. Kilby, Tom Carr, Herbert Braren, and me. We located our camp on the beach at the south end of Seahorse Island, which lies about five miles northwest of Cedar Key.

Seahorse is a roughly crescent-shaped, fairly well-wooded island that is two or three miles long. It has a large population of hogs, a boarded cistern for them to wallow in, and an abandoned lighthouse on its highest point.

We arrived rather late and set about making preparations for retiring. Kilby, objecting to the arenaceous nature of the communal couch, retired to a distance of fifty yards or so back of the beach, where the grass was thicker. Suddenly we heard shocking language in his quarter, and he emerged in great haste, shouting that he had laid his blankets on two adult cottonmouth moccasins and dragging the mutilated corpse of one of them to support his story.

Stimulated by this experience, the party dispersed to explore the island. Within an hour three more moccasins were discovered. One of them was coiled at the base of a cabbage palm near the beach; the other two were nearly trodden upon in the trail leading from the beach up to the lighthouse.

The following morning we set out to investigate the validity of the name Snake Key, as applied to another little island four miles off the mainland to the south of Seahorse. We found it to be a narrow strip of land about a mile long and a quarter of a mile wide, bordered along two sides with a thick growth of red mangrove. Inside the island we were surprised to discover a well-developed forest of shore bay *Tamala littoralis,* with scattered cabbage palms and little undergrowth other than an occasional patch of wild pepper bushes. The ground is overlaid by a thick carpet of bay leaves, and the sunlight, filtered and broken by the interlocking limbs above, falls in a pleasing mosaic on the forest floor. The whole aspect is very reminiscent of a high hammock on the mainland.

As we sauntered through one of the long aisles among the trees, I happened to direct my glance downward. There at my feet was a young cottonmouth, neatly coiled, his pattern blending perfectly with the chiaroscuro of the background. I impaled him with a thrust of a frog gig that I was carrying. Turning to exhibit my capture to Bellamy, six feet behind me, I again looked down. To my alarm I perceived a broad black head belonging to a body hidden under the leaves directly in Bellamy's path, where he could not fail to step on it. Because his feet were clad only in tennis shoes, and because the two steps that would place him squarely over the snake were being executed with energy, I made recourse to the only means of stopping him that I could conceive on the instant—I jabbed him viciously with the gig, adorned though it was with the still-living cottonmouth. Bellamy was justly outraged at the act, and Kilby and Tom, who brought up the rear, regarded the scene with grieved astonishment. As I pointed out the cause for my show of violence, the latter two suddenly uttered cries of warning and scaled a nearby tree with great alacrity. From a branch they indicated that a third moccasin lay a few feet away, nearly covered by leaves.

After a brief period devoted to recovering a semblance of composure, we bagged the three snakes and resumed our stroll.

During the course of our traversal of the island we caught ten more cottonmouths. We didn't go out of our way to search for them—we merely tried to avoid stepping on them. It was with some relief that we reached the opposite end of the island. We returned to the boat by the way of the beach.

It is difficult to account for the presence of such a tremendous cottonmouth population in a situation of this nature. We found no trace of fresh water on the island. The only other terrestrial vertebrate that we encountered was the Florida five-lined skink (*Eumeces inexpectatus*), which was fairly abundant among the dead leaves in the woods. The bay trees harbored a large number of wading birds, most of which were nesting; we identified the following species: Ward's, little blue, little green, snowy, Louisiana, and black-crowned night herons. We saw no sign of rabbits, rats, or other small mammals, and terrestrial birds were very scarce. Apparently then, the food sources available to the snakes are three in number: the heron rookery, the skink colony, and the marine fish population.

The herons are there for only a short period of each year; even then, the most to be expected from them is the occasional toppling out of the nest of an egg or fledgling. The skinks, though perennial inhabitants, are small, nimble, and apparently not much more numerous than the snakes. Saltwater fish are plentiful enough, but it is difficult for me to envision a cottonmouth pursuing its prey in the open Gulf or in a mangrove swamp. The improbability of the occurrence of this perversion is supported by our failure to encounter even one of the snakes near the water, or in fact anywhere except in the dry woods in the interior, even though on several occasions we walked around the island and through the mangove swamp.

The possibility of temporary, seasonal, or sporadic occupancy of the island by the moccasins seems to me very remote. I have seen other terrestrial and freshwater snakes in salt water—on two occasions, rattlesnakes—but I never saw or heard of a cottonmouth voluntarily taking to the sea.

An account of the stomach contents of the thirteen snakes taken on the island is presented here. The snakes have been identified by age and sex.

1. yearling: none

2. adult female: three heron feathers

3. adult male: heron feathers

4. young female: none

5. adult male: bird bones

6. adult female: bird bones

7. young female: one skink (*Eumeces inexpectatus*)

8. adult male:none

9. adult female: one skink (*Eumeces inexpectatus*); three fish, all under one inch in length; one heron egg shell

10. adult female: none

11. yearling: none

12. adult female: none

13. young male: none

It will be noted that the most salient general feature of the stomachs is their vacuity.

The most interesting item in the list is the three small fish. At four feet, ten inches, No. 9 was the biggest snake we caught. The fish were very small: two of them were three-quarters of an inch long, and the third was less than a half-inch in length. The necessity for believing that this massive serpent had engulfed these tiny fry with the aid of dental equipment too heavy for prey five times as big was disturbing. It was with relief that I finally realized that I had picked the fish, with forceps, out of an eggshell and had lain them aside for identification. Sensing a way around the impasse, I returned to the eggshell and inspected it carefully. There to my satisfaction I found on its inner wall two spots of white guano and a streak of dried mud. The snake was vindicated. She had not eaten the fish at all, but merely an old eggshell with the smell or taste or aura of fish and bird about it. The mother heron had caught the fish, and the sloppy youngsters had spilled them into the eggshell in the bottom of the nest. At a subsequent housecleaning the shell had been ejected. Mingled with my satisfaction at the neat deduction was a feeling of pity for the cottonmouth. How the shell was ingested without being crushed to bits I cannot imagine.

Two observations that I find in my notes on the moccasin hunt impress me as being of such an esoteric nature that mention is made of them here with the greatest trepidation. I record them only as statistical facts, with the assurance that I have conceived no explanation for them.

Of the thirteen snakes encountered, five were young, with the juvenile pattern of alternating wide bands of brown and gray; the remaining eight were old individuals of uniform black coloration. The young ones were all found lying in the open on top of the leaves, where they presented the most remarkable example of protective coloration that I have ever seen. Two of the eight black ones were crawling over the ground, but the other six were without exception coiled beneath the leaf mold, with only the head protruding.

Further, out of the thirteen snakes, all but the two that were moving about were located under trees in which there were heron nests.

I leave to your discretion speculations on whether and how the young snakes knew they were protectively colored and the old ones that they were not. Moreover, I disclaim all responsibility for their lying under the nesting trees, and I know no more than you whether or how they knew they were under nests and that sooner or later an egg or a young bird or a fish would fall out.

In fact, there is little about these island cottonmouths that I do understand. But seven months later, during their breeding season, we came upon a three-foot male and a monstrous female whose old skin had broken away from her lips and head and stood out around her neck like a Queen Anne collar. She started gliding away at our approach, but her consort, lying patiently by her side, ignored our presence and, gaping his fearful mouth, seized her gently about the middle and detained her. That, I believe, we can all understand.

And I also know that the problems presented by the Snake Key moccasins are fundamental ones that, for personal and biological interest, would more than justify the time spent in their solution.

A *D*ubious *F*uture

If the world goes on the way it is going, it will one day be a world without reptiles. Some people will accept this calmly, but I mistrust the prospect. Reptiles are a part of the old wilderness of Earth, the environment in which man got the nerves and hormones that make him human. If we let the reptile go it is a sign we are ready to let all wilderness go. When that happens we shall no longer be exactly human.

One of the awesome enigmas of today is how to slow the ruin of the natural earth while our breeding continues. There is no more need to multiply with the old fever. Breeding is good business, but it is herding our race toward a tragic impasse. When this is clearly seen and the reproduction is slowed down it will be because thoughtful people have taken charge; and these people will look about for what has been left of old values. One of the values is what the human spirit gets from wilderness—from all kinds of wild original landscapes and beings. The way we are going, what we keep of the old Earth will not be enough to save our honor with our descendants.

Writing this, I felt one of the qualms you cannot keep down when in your mind you weigh new industries against rough country empty of all but unused beasts and vegetables. I have no real doubts myself, mind you, but to many others in the world, especially the Florida world, to question the complete goodness of population growth is a perverse and sinister sort of iconoclasm that probably should be investigated by a committee. Thinking that way, I scared myself a little, and to get over it I called off the writing for a spell and went over to Lake Alice. Lake Alice is one of the solid assets of the University of Florida.[1] It is a sinkhole lake with tree-swamp at one end and open water at the other, and all through it a grand confusion of marsh creatures and floating and emergent plants. The place is a little relic of a vanishing past, and, incredibly, it lies on the campus of a university

---

Adapted from *Life Nature Library: The Reptiles,* by Archie Carr and the Editors of Time-Life Books. Copyright © 1963 Time-Life Books, Inc.

with thirteen thousand students and less than half a mile from where I am writing now. It is there to go to when euphoria spreads through the press over some new gain the state has made in people.

I went this time to where an alligator called Crooked-Jaw has her nest beside a wire fence at one edge of the swampy end of the lake. I stopped the car and walked over to the nest and looked at it closely. I had taken a picture of it the day before, and I could see that Crooked-Jaw had made some changes during the night. They were not drastic—only small, fastidious adjustments to show she knew the heap was warming a new generation of her kind. A root-mass of buttonbush had been added, along with a few live switches of *Decadon* and some scooped-up slush of coontail from the bottom. On the top of the pile was a single balled-up pink paper towel; and though it seems unlikely, I am sure I had seen this lying six feet to one side of the nest the day before. I can say that because I was aroused at the sight of it, at the idea of anybody defiling with pink the premises of an alligator nest. Crooked-Jaw clearly failed to share my resentment. There is no accounting for tastes. The nest did not look as good to me with the paper towel on it, but the matter was not in my hands.

The alligator was not in her usual station, her lying-in pool, as it were—the little dredged-out hole of water a mother gator waits in for eggs to hatch. She was off somewhere among the floating islands, and I started croaking—*eer-rump, eer-rump*—like a little gator. A long way out through the flooded willows a floating island began to quake; and then all at once water surged out from the frogbit raft beside the waiting pool, and Crooked-Jaw came up looking at me. A gallinule whined from a bonnet patch, and in the high haze to the west, the sandhill cranes were bugling. I croaked some more, but the alligator had lost interest. She sank into the water till her chin rested on the mud, and only the bumps of her eyes and nose and the big scales of her back stuck out.

Looking at her there in her fragment of a doomed landscape, I was sure again that the saving of parts of the primitive earth has got to be done, and that it has got to be done without trying to justify it on practical grounds. Species and landscapes must be kept because it pleases people to contemplate them and because freer men of future times will be appalled if we irresponsibly let them go. Not facing that fact seems to me the great weakness in the outlook for wilderness preservation today.

It will take resolute people to put abstract values in place of material progress. In testing the mettle and conscience of recruits for the work, the reptile—particularly the unloved, legless snake—may serve as a sort of shibboleth. A man who feels in his bones that snakes must be kept in the woods will be proper stuff for the struggle coming.

Snakes are probably disappearing at a more rapidly rising rate than any other group of vertebrates. Besides the widespread antipathy they get from man, marshes are drained, country is reforested in pure stands of unsuitable cover, poisons spread abroad kill off the food supplies of the creatures snakes eat and even kill the snakes themselves. But the most spectacular thing happening to snakes is the onslaught of cars on the roads. In his book *That Vanishing Eden,* Thomas Barbour spoke of the passing of snakes before cars on the roads of Florida, but he never saw the big change. It came with the many-laned highways of the fifties and sixties.

The worst snake traps are the causeways across marshes and the streams of cars that cross them. Snakes are lured to them to enjoy the warm pavement or to escape flooded habitat, or they encounter them merely in the course of their foraging. I remember a vast dying of snakes on the road across Paynes Prairie decades ago, when man and weather chanced to move together against the creatures of the marsh. On October 18, 1941, a hurricane moved in from the Gulf and spun in the vicinity for thirty-six hours, bringing fourteen inches of rain during five days. The prairie changed from a marsh to a lake, and the water rose so high that only the tips of the tallest grasses showed. On the twenty-fifth some students brought in two hundred snakes they had caught along the road-fill and told of a great hegira of snakes and of congregations of buzzards squabbling over the ones mashed by passing cars. There was clearly something extraordinary going on, and four of us from the biology department went out to investigate. We started at the northern edge of the prairie and walked abreast down the road with flashlights, one of us at each guardrail and two along the middle of the pavement. The road over the marsh was two miles long. We counted every snake dead or alive between the guardrails, which in those single-lane days were twenty feet apart. We picked up 723 snakes in the two miles, about two-thirds of them dead or injured.

As an accumulation of several days, this number of casualties would not have been unprecedented. But these were the accumulation of no more than the four hours or so since sundown. During the daylight hours buzzards—black vultures and turkey vultures—had been attracted to the killing by the hundreds and had carried the dead snakes away almost as fast as they were run over. So the snakes we counted had been killed after dark. The tally was: 284 red-bellied snakes; 200 ribbon snakes; 85 green water snakes; 64 banded water snakes; 55 garter snakes; 19 Allen's mud snakes; 6 brown snakes; 4 cottonmouth moccasins; 3 horn snakes; 3 king snakes.

The slaughter had no noticeable effect on the levels of snake populations in the prairie. For a decade afterward the road remained a mecca for snake collectors, and they kept coming from distant places to walk along it with bag and stick. But in recent years the prairie snakes have declined. Although the roadside was made a wildlife sanctuary, and the snakes in it are now immune to people who used to take them away in sacks, the cars keep going by, and snakes have no immunity to them.

No significant preserving of nature can be done with slight sacrifice. The true test will come when great sacrifices are needed, when it becomes necessary to fight the indifference of most of the world and the active opposition of much of it, to surmount man's ingrained determination to put the far future out of his mind in matters of current profit.

Besides the inherent technical difficulties of wilderness conservation, the effort to save original nature faces a whole constellation of other kinds of problems. The easiest obstacle to recognize is the opposition by people who for material reasons oppose the keeping of wilderness. There is another block of humanity that simply does not care and an unsorted lot made up of those who think of themselves as conservationists—and who in one way or another are, but who are not facing the really tough obligation at all. I refer to all people who think of saving nature for meat, water, timber, or picnic grounds for the future; and to the hunters who hope their grandsons will get red blood by shooting things; and to the reverence-for-life cultists who are foredoomed to inconsistency; and to the biologists who resist the loss of material for study; and to keepers of zoological gardens who preserve nature in cages. Putting this mixture of motives and aspirations together under the label conservation has made, in some cases, a temporarily stronger front. But it has muddied the real issue, hidden the dimensions of the long job and kept everybody from articulating the awful certainty that the hard saving has got to be done for the sake of abstract values.

For several years I have been involved in a preservation program that has been atypically feasible. This is a campaign to rehabilitate the green sea turtle, *Chelonia mydas,* in the Caribbean Sea, where its once extensive nesting range has been reduced to only two rookery beaches.

In the Caribbean, the way things were going a short while ago, the green turtle was facing complete extirpation. Now I believe there is no such danger. The change in outlook was made possible by a combination of circumstances such as cannot be counted on in most preservation projects. In the first place, the suspected migratory feats of green turtles focused scientific interest on them and brought support from research foundations—the National Science Foundation and the Office of Naval Research—for studies of their basic natural history. A major factor that has greatly eased the way for preservation is the lucky circumstance that the single nesting beach remaining in the western Caribbean is located at Tortuguero, on the coast of that gem of a small nation, the Republic of Costa Rica. In former times exploitation of the Tortuguero colony brought Costa Rica a steady small revenue in the form of a fee paid by the concessionaire, who parceled out the beach to the turtlers and sold their catch to the Cayman schooners or sent it away as deck-loads on freight boats going back to Florida. But in 1957 the government closed the beach to exploitation. The move saved the green turtle for the western Caribbean, but it also deprived Costa Rica of all profit from its green turtles because there is no good turtle pasture along the Costa Rican shore and no turtles go there except during the breeding season. The refuge will repopulate the pastures from Colombia to Mexico and will increase the yield of the turtle grounds of the Nicaraguan Miskito Coast to schooners turtling for the markets of New York and Europe. For Costa Rica itself there is only the satisfaction of having faced the choice between quick gain and a better future—and having chosen with characteristic wisdom.

In 1955, when the first of a series of grants from the National Science Foundation was made, a tagging camp was established at Tortuguero. The information accumulated helped stimulate the founding of the Caribbean Conservation Corporation, a nonprofit undertaking dedicated to restoring the Atlantic green turtle in American waters.[2]

The world is responsible for reptiles. The inadvertent saving of scraps will never keep off the ruin of the earth. The only way is to name the real obligation clearly, to say without hedging that no price can be set for the things that have to be preserved. Basically, what must be done are the harder jobs, like justifying a future for snakes, which have no legs, hear no music, and badly clutter subdivisions. Bore through to the core of what is required and you see that it is an aggressive stewardship of relics, of samples of original order, of objects and organizations of cosmic craft. This work will take staunch people, and the reptile can be the shibboleth by which they pass.

To get the real feel of the problem, I conjure up a man of some far future time walking in a last woods lying unruined among launching pads of a planetary missile terminal and coming astounded upon the last of all living individuals of *Crotalus adamanteus,* the great unruly diamondback rattlesnake. It is a full-grown female snake that I see, two yards long, stern of face, and all marked off in geometric velvet. It is the sort of being that always, inadvertently and without malice, has been a thorn in the flesh of Americans, one of the novel terrors the land held for humans whether they came in caravels or wandered down into the New World out of the snake-free Siberian cold. Seeing the man, this last diamondback begins readying the steel of its coils, and they ebb and flow behind the thin neck holding the broad head steady and still, except for the long tongue waving. By the girth of her I judge that this is a pregnant snake, heavy with some dozens of prehatched perfect little snakes the same as herself, all venomous and indignant from the start, all intractable and, like their mother, unable to live except as free snakes.

The snake that confronts the imagined man is a moving thing to see. It is not easy to understand all the feelings aroused by such a sight, and the snake I think forward to is the last in all the pabulum agar culture of the purified world. The coils of her body rise and fall in slow spirals, the keen singing of her rattle sounds, and she waits there, testing with the forks of her tongue the whole future of her kind. In my thought the man then stoops with an old urge and picks up a stick. It is almost the only stick left lying in the eastern half of North America, and the man takes it up and moves in closer to the wondering snake. He raises the stick, then somehow lowers it as if in thought, then halfway brings it up again. And then the conjuring fails for me, and the snake song falls away, like the song of cicadas losing heart, one by one. The woods grow dark and fade off into distant times.

*E*den *C*hanges

The history of man and nature in Florida has not been a wholly happy one. Very recently it became possible to balance some of the violence we have done to the natural world with a few signs of goodwill; but the change was slow in coming, and the delay has cost us heavily.

The irony of this is that, more than almost anywhere else in America, it was nature that drew people to Florida to start with. Partly the early violations were a sign of the times. To our forebears, cleared land was better than forests, and killing nonhuman creatures was only natural. Besides that there was from the beginning an obsession with the goodness of population growth. From the outset the natives saw profit in visitors who could be overcharged for a side of bacon or a mule, or guided to good shooting in a heron rookery. This helped generate the heady magic of growth and made the word synonymous with progress.

The history of Florida has been a desperate sort of striving for growth and development. The result of this has been the most protracted crowding in of outsiders that any state has had. Inevitably, this has dimmed or destroyed much of the natural charm that originally drew people here. There is no way that the favors of wild nature can be infinitely shared about, like loaves and fishes. The climate has not so far been materially hurt, but all other ecologic assets have to some extent been changed, and some have been forever lost. But it is no longer a walkaway for the fast-money chaps, and though they still have a lot to learn, they seem to be learning it.

When a naturalist whose delight is wild creatures and unworn country sets out to write about Florida, it is hard to refrain from bemoaning lost wilderness, as John Small did in *From Eden to Sahara* or Thomas Barbour did in *That Vanishing Eden*.[1] Actually, if you look around for them, there are pleasant things to be told. And there are things to be thankful for. The return of the manatees to the limestone springs, for instance. Thirty years

From Joan E. Gill and Beth R. Read, eds., *Born of the Sun: The Official Florida Bicentennial Commemorative Book* (1975).

ago it was next to impossible to see a manatee in the primordial wintertime habitat of the species—the big springs of the central peninsula. Manatees are good to eat, and they were hunted out of the springs long ago. Those not killed for meat were idly shot by duck hunters or by kids with .22s. Today the poachers and irresponsible gun-toters are fewer; and manatees, though still an endangered species, are back, in some of the springs at least. And that is a blessing because manatees are neat, very neat.

It is somewhat the same with alligators. When Bartram was here, there were alligators galore. By the late 1800s, however, hide hunting and recreational slaughter were going on everywhere, and alligators had disappeared from much of their natural range. Populations declined drastically, even in Florida, and tourists could count on seeing alligators only in the alligator farms. A few years ago the alligator was declared an endangered species. Under protection it has proved surprisingly resilient. In a few parts of its original range the species is obviously no longer endangered. A visitor who looks around can now usually see wild alligators in natural habitat; and in some suburban bodies of water they have even become a nuisance. The existence of these localized sites of abundant, brash alligators generated the retrogressive scheme to reopen a hunting season. The potential for harm in this proposal is complex, and one can only hope that nothing comes of it. Meantime, having visible wild alligators in the landscape is a thing to be thankful for.

Besides that, there are otters. During the first half of this century otters were heavily trapped and shot, and seeing one became a rare occurrence. Today hunting pressure has relaxed, and the main enemy of otters is the highway. Otters are very bright, but for some reason their brain copes poorly with automobiles; they are often found dead on any stretch of paved highway through marsh country. Away from the fast cars, though, they are much more numerous now than they were in the 1940s, and their return is cause for celebration.

So is the return of the beavers. Beavers probably never entirely disappeared from the Panhandle of Florida, but as far as most Floridians could tell they might as well have been completely gone. Now they are increasing their range and abundance in the state. A few single-minded entrepreneurs have even begun to cry out for beaver control on the grounds that timberland is being flooded by their work. But most people probably would figure beavers in Florida are worth a few drowned planted pines.

Another pleasant change—one more conspicuous to the casual visitor than the return of sea cows, alligators, otters, and beavers—is the spread of long-legged wading birds. A part of the new look that these produce is contributed by the cattle egret, an Old World species that for some wholly unknown reason began crossing the ocean three decades ago and is now a common sight standing with cows in pasture lands. But these new immigrant white egrets being here should not obscure the important fact that other water birds too have returned as a regular feature of the landscape, in the roadside ditches and out on the wet prairies. It took many years for them to recover from the plume-hunting massacres that were finally stopped by the Audubon Society wardens in the early 1900s. But herons are again decorating the wet places. Even roseate spoonbills can be seen with little searching, and this was undreamed of only a while ago.

There has been another pleasing change in the look of much of the landscape of interior Florida. Open country free of people and agriculture is more lush and comforting to the eye than it used to be. This change has come about because of the decline of the Cracker habit of letting wildfire loose in any patch of woods any time it would burn. Planned, monitored burning of pinelands is a useful tool for keeping out hardwood seedlings that would turn pine woods into hammock. In fact, it is altogether essential to the maintenance of pine flatwoods and of other pineland as well. But unless the burning is done with skill it makes a shocking mess of the land. The old Floridians burned heedlessly, aiming only to be rid of debris from timbering or to stimulate the growth of new grass that would rescue their cattle from a winter diet of palmetto fans. Or, if they had no cows, they burned to kill ticks and rattlesnakes or to flush out game.

Some people just don't like the look of flatwoods anyway. Jacob Rhett Motte, an army surgeon of Charleston, South Carolina, who arrived in Florida during the Second Seminole War, was appalled by the appearance of the pinelands around the Georgia-Florida line. They seemed to him a "dull, insipid pine barren, where the listlessness of blank vacuity hung upon the flagging spirits."[2] Motte was probably just prejudiced. Most naturalists would give a lot for a chance to see what those pinelands looked like before they were cut down. But later on the flatwoods were cut over, most of them more than once, and the Crackers burned them into blackened, bare-floored semidesert, devoid of animal life except for razorback hogs that plowed up the bayheads and thin sooty cattle that stood around in the road or wandered through the unfenced country foraging morosely for scattered sprigs of wire grass.

Now the ritual burning has stopped. In some cases the protection is overdone, but a new and pleasantly lush look has come over the land. It is not a wholly natural look. Even before the Crackers and Seminoles came to Florida the woods used to burn over, in natural fires, and the present fire-free regimen will make hardwood hammock of much of what once was pineland. Meantime, however, what you see from a traveling car is far greener and more opulent country than what they used to see from the Model T Fords.

So there are bright spots in our relationship with natural Florida. When you think of the abuse much of our heritage has suffered, however, it is not easy to stay cheerful. Take the decline of the big springs, for instance—the biological communities in the limestone springs of the state. The ecological degradation of these incomparable springs is one of the major losses the state has suffered and one that clearly illustrates the problem of saving the diverse values of natural landscapes.

Florida is underlain by soluble limestone. Where this lies near the surface, solution has formed caves, chimneys, and sinkholes, and much of the drainage of the region goes into subterranean streams rather than running over the surface to the sea. Rivers disappear into the ground and emerge a mile or more away; lakes gurgle out overnight through newly opened holes in their bottoms. Such terrain is known as karst topography, after a region in Hungary. There are numerous karst regions in the world, but nowhere are big river-making springs as abundant as they are in Florida.

And nowhere in the Florida landscape is natural beauty distilled to its essence as it is in the big springs—wherever their natural biologic organization has been spared. There is a dreamlike quality to the appeal of these places. It is a stirring thing to come upon a line of them unexpectedly, whether you walk in overland and find it suddenly glowing in shadows of live oaks and magnolias or you paddle up a cypress-bordered run, wondering where the run comes from, and then all at once see the trees open in a circle and live water surging up like liquid blue crystal. When William Bartram wrote of his journeys through the limestone country in the late 1700s the springs evoked some of his most rhapsodic prose. Three hundred years before that, it was surely rumors of the supernatural beauty of the springs that generated tales of the Fountain of Youth.

Each spring is different from all the others; but in the intensity of its grace and color each is a little ecologic jewel in which geology and biology have created a masterwork of natural art. In all of them the water wells up in shades of blue and silver that vary with depth and the slant of your view. In the unviolated basins there are submarine beds of water plants of half a dozen kinds, each a different shade of red-brown and green waving slowly in the current or spreading over quiet bottom like patchwork quilts of velvet. Where the flow is too fast for plants to hold, the bare sand of the bottom shows white, and in some of the springs white chips of fossil shell or flakes of marl swirl up like snow with the roiling newborn water.

It would be hard to find a better example than these springs to show the kinds of troubles that hinder the preservation of biological landscape. The springs are still there, most of them. Fountains still surge up out of rock caverns and make sudden streams that give a strange look to the contour map. But all of them are distressingly fragile treasures that have without exception suffered damage, in some cases irretrievably.

One of the magnificent springs of the Gulf drainage of Florida is Manatee Springs, which makes a tributary to the Suwannee River in Dixie County. It was once an old haunt of mine. I used to go there, partly because two species of freshwater turtles I was interested in lived thereabouts, but also simply because it was a lonely, magic bit of landscape. I first saw it in 1935. There was little animal life of any kind within the spring itself except for occasional roving schools of mullet; a trio or pair of needlefish here and there; a few red-eyed Suwannee bass; itinerant small gangs of bluegills, redbreasts, punkinseeds, and stumpknockers; a few shy Suwannee chicken cooters, one of the turtles I used to go there after; and great numbers of big-headed stinkjims, known less colloquially as loggerhead musk turtles.

That was long before the spring had become a state park. The road out there from Old Town was a long sand-track through the flatwoods. The place was so hard to get to that hardly anybody ever went there except to make moonshine or to dynamite fish. The sparseness of the fauna of those days was no doubt mainly the work of the dynamiters—the cut-bait fishermen, as they are euphemistically called.[3]

And I knew one of the best of all the cut-bait fishermen. His name was John Henry. He used to bring his mule to help Tom Barbour and Ted White dig out fossils at the Thomas Farm over in Gilchrist County northwest of Bell. The Thomas Farm dig had just become known to paleontologists and was being called the best deposit of Miocene fossils east of the Mississippi—or of the Rockies, I forget which. Anyway, Dr. Barbour bought the place for Harvard and the University of Florida and used to come down once in a while to help Ted work it. My wife and I used to go out there with them. They would work away with a grapefruit knife for several days, scratching the clay away from the skulls of horses or camels or dier-wolves that died thirty million years ago, and then, when everything the grapefruit knife could reach was out and safely shrouded in plaster, John Henry would bring over his mule and mule-shovel and scrape off the overburden until more pay dirt lay within reach of the grapefruit knives. There were two other local fellows who came out into the worn-out turkey-oak farm to help with the digging, but they weren't cut-bait men. One was called Uncle Leonard, and he was a solid, quiet man. The other was known to me only as Uncle Goo; I don't remember why. It could have been because his teeth were all gone on one side of his lower jaw, and snuff juice kept running out that side of his mouth and down his chin. He never bothered to wipe it away. Anyway, I can't help but take a moment to tell about the time they were all out there watching Ted tease away the final scraps of clay matrix from around the skull of a giant Miocene carnivore of some kind—a bear-dog, Ted called it, a creature of the line from which all the bears and dogs of today have come from, and with canine teeth more frightening than you ever saw. The three local men had been at work for a couple of seasons and had grown fairly familiar with the fauna of the dig. They gave their own homely names to the fossil bones that appeared, as well as to the animals the bones suggested to their minds. A femur of a giant peccary was a "hambone of the boar-hog," for example; and they had a long list of other terms relating the bones of animals nowadays to those of species no longer prevalent in Gilchrist County. One lazy afternoon, while waiting around for Ted to get the bear-dog out of the clay, they idly fell to wondering about the times when such giants were in the land, and Uncle Leonard said, "You reckon there was any folks around here in them days?"

"Shore," John Henry said. "There's always been folks around these parts."

Ted was used to the way his colleagues telescoped time in order to make sense out of the incredible beasts they dug out of the dig, so he just kept on scraping out clay from around the emerging great jaw and said nothing. Leonard pondered the faraway times for a while, and then he said, "Well, what kind a rig you reckon they had to handle a critter like this-yer-un?"

John Henry faced him and scornfully spit snuff juice against a clod of blue clay. "Great God-a-mighty," he said. "What you reckon they had them long roffles fer?"

Leonard nodded thoughtfully and said, "Shore," and Goo said shore, too.

That digression had nothing to do with my subject, except that besides knowing answers like that, John Henry was the champion cut-bait fisherman between Bell and Branford. He knew how to drop a weighted half-stick of "powder," as he called dynamite, and turn a single good channel-cat belly-up twenty feet down in a spring. He could fix up a charge with just the proper fuse and weight, drop it down into a dark eddy of the Suwannee and get out with enough fish for a church fish fry long before the game warden came sneaking down the river to where he had heard the deep bump of the explosion that meant cut-bait fishermen were at work. The best test of John Henry's skill was cut-baiting for mullet. He would take a little chunk of powder, a quarter or an eighth of a stick, and put on a cap and a fuse so short that lighting it seemed like suicide; and he could get that out into a nervous school of big, cruising mullet in an arc so well-timed that it would go off as it hit and blow the fish out of the water before they could shy at the splash.

John Henry was proud of his skill with dynamite. He kept some of it on him most of the time. According to Goo, he had some on him one time when they put him in jail for fighting. They overlooked it in his pocket, and John Henry blew a wall out of the jail and went away. Nothing ever came of it, either, Goo said. What John Henry said at the time was that somebody else blasted the wall from the outside.

Although my digression got out of hand, it has a relevant point, really, which is that the cut-bait fishermen, the dynamiters of fish, used to be dismally prevalent in Florida, and they undoubtedly routed great quantities of fish out of many a spring.

At Jody's Spring, near Silver Glen in the Ocala National Forest, there is a charming bit of the world, described by Marjorie Kinnan Rawlings in the opening chapter of her book *The Yearling*.[4] It is a place where you can sit for a while in the dim cool of a scrap of hammock surrounded by sand-pine scrub and marvel over a superb small gem of the natural landscape.

Jody's Spring is unique. There is no one big, river-making outpouring but instead a scattering of gentle little geysers of crystal water and snowy sand bubbling in the bottom of a shallow pool surrounded by evergreen hammock. Each sand-boil makes a lively snow-white pit in the leaf-strewn bottom. Some of the boils are no bigger than your fist, some are the size of a washtub. Killifish cruise about among them, and wherever one is big enough to hold him, a young bass or half-grown redbreast bream is usually ensconced, working his gill flaps and eyeing any visitor through the air-clear water. Thin conical snails creep across the brown leaves on the bottom of the pool; and dragonflies course above it or bask on twig-tips in splashes of sunlight. Where the outlet leaves there are patches of cress and lizard's tail, and neverwets spread velvet leaves about the banks, their flowers glowing gold in the gloom. The pool feeds a stream that wanders away through the woods to join Silver Glen Run a short way off. In any setting Jody's Spring would be an enchanted place. Set out there in the heart of the vast hot scrub, with little bubbling boils gleaming white and silver in their quiet patch of deep, cool shade, a marvelously unreal aura is generated.

There are ecologic lessons to be learned at Jody's Spring. The hammock is a striking variant of the scrub community where, because of the presence of the springs, the sand-pine forest gives way to a moist woods of broadleafed trees. There is an abrupt transition, a narrow ecozone, where tall cabbage palms can be seen standing almost side by side with the closely related dwarf-sabals that grow only in scrub. And yet if you go there and sit in the cool quiet for a while, the important thing you will see is a work of art.

As you sit there admiring it, however, as likely as not a car will come tearing up and stop out on the road. Doors will slam and people will charge down through the hammock, thrash out into the pool, and enter into a raucous competition to see who can sink down deepest into the heaving sand of the little springs. To a quiet watcher on the bank the invasion seems a violent assault. The worst of it is, to the people engaged in the assault it is harmless horseplay and a lot of fun—an offbeat and stimulating thing to do. So the predicament of Florida's springs clearly exemplifies the intractability of wilderness preservation when both aesthetic values and opportunities for physical recreation are involved—which they almost always are.

This dilemma hinders most efforts to save wild places to which the public is admitted. In the case of the springs, their unique value is a fragile loveliness that depends on their integrity as biologic landscape. They are all attractive for other reasons, too; and many of the people drawn by these other qualities often miss the real point completely.

Long before Florida settled up so badly, the springs had begun to suffer. Those located in farmland were used for irrigation. The moonshiners liked the water for their stills. Springs near towns made superb swimming holes, and some of them became popular spas, to which people from other parts of the South came every summer to take the waters. So the wear and tear began long ago, but more recently it became much worse. Madmen in outboard-powered boats have raced round and round in the boils, making deserts of the basins. And into the deepest and most enchanted the scuba divers have gone. They come from everywhere, by the hundreds, to test their skills in enchanted caves and fountains, and by the mere passing of their countless bodies and the bubbling of their regulators scour and scare to lifelessness some of the best of the spring communities.

With most of the Florida rheocrene springs already damaged and some utterly wrecked, any that remain in anything like natural ecological diversity and organization ought to be made inviolate sanctuaries, kept perpetually free of contact with either boats or human bodies. That is tough, because a part of the artistic appreciation of springs is getting into them, putting on a face mask and going down and looking through airlike water at nuances of light, life, and color never thought of back on the bank. One human quietly flippering about in a spring or spring run does no harm to speak of. But as viewers multiply, even reverent ones, the place begins to wear. So there is really a cruel dilemma to be faced, if even the handful of unspoiled springs is to be saved. And while the trouble reaches a peak in the special case of the springs, it is much the same wherever the complex organized interplay of animals, plants, and their living space is the treasure to be preserved.

In listing some reasons for optimism over the state of nature and man in Florida, one favorable development outweighs all the rest. It is not another species on the mend or a new park or preserve or sanctuary established. It is rather a change in the heart of the people. Although original Florida is still undergoing degradation, an assessment of the trends would show the rate of loss being overtaken by the growth of a system of ecologic ethics, by a new public consciousness and conscience. There was a time when "preservationism" was a dirty word, a name for visionary folk whose aim was to keep the world the way the Indians had it. But now the farsighted kinds of people who saved the white birds in 1913 and three decades later generated the Everglades National Park, have multiplied and are influencing the whole political climate for conservation and preservation. In 1972, by a 65 percent majority, Floridians voted to tax themselves for an endangered lands program, set up to purchase outright wild land threatened by development.[5] These changes in the public mood are reflected in government policies as well.

The rise of this new stewardship gives heart to opponents of ecologic ruin everywhere and brings promise of better times for man and nature in Florida.

Notes

Prepared by Marjorie Harris Carr

Jubilee

1 Harold Loesch, "Sporadic mass shoreward migrations of demersal fish and crustaceans in Mobile Bay, Alabama," *Ecology* 41 (April 1960): 292–98.

2 Archie Fairly Carr, Jr., *A Contribution to the Herpetology of Florida*, Biological Science Series, no. 3 (Gainesville: University of Florida, 1940).

3 William Bartram, *Travels Through North and South Carolina, Georgia, East and West Florida* (Philadelphia: James and Johnson, 1791; New York: Penguin, 1988). See also Francis Harper, *The Travels of William Bartram* (New Haven: Yale University Press, 1958), 120.

4 Homosassa Springs is today the centerpiece of the Homosassa Springs State Wildlife Park—155 acres of some of Florida's loveliest landscapes, including marshes, swamps, hammocks, and spring runs. For many years the springs were privately owned and operated as a high-class tourist attraction. A floating underwater observatory allowed visitors to watch the myriad fish at close range. The state of Florida purchased Homosassa Springs in 1989 and continues to operate the observatory. The objective of the park is to provide a showcase for native Florida wildlife and endangered species. For more information contact Homosassa Springs State Wildlife Park, 9225 West Fishbowl Drive, Homosassa Springs, FL 32646; telephone, (904) 628–2311.

5 Since 1977 the St. Johns River has been under the care of the St. Johns River Water Management District, one of five water management districts in the state. Each district has taxing authority, so funds are available for innovative restoration, as well as maintenance. Years ago the headwaters of the St. Johns were drained for agricultural use. Now nearly 200,000 acres have been purchased and returned to the original marshland state. In addition, the district has undertaken the restoration of the Ocklawaha River, the largest tributary of the St. Johns. Lake Apopka, one of the headwater lakes of the Ocklawaha, had been in an advanced state of eutophication for many years; now it is rejuvenating. The floodplain of the upper Ocklawaha had been drained and dredged, and sections of the river were channelized. These mucklands are now being bought up and returned to the marshlands characteristic of the floodplain. Restoration is slow process—and the slowest part is getting people to take the initial step. Once nature is given a free hand, the process takes place with startling rapidity. For more information contact John Hankinson, Director of Planning and Land Acquisition, St. Johns River Water Management District, Box 1429, Palatka, FL 32077; telephone, (904) 329–4500.

Sticky Heels

1 Archie Carr, *Handbook of Turtles: The Turtles of the United States, Canada, and Baja California* (Ithaca, N.Y.: Cornell University Press, 1952). Archie started work on this book in 1944 before we moved to Honduras for a four-year stint with the Escuela Agricola Panamericana, an agricultural school for Central American boys that was sponsored by the United Fruit Company. He completed the book after we returned to Gainesville in 1949.

2 J. T. Nichols, "Data on size, growth and age in the box turtle, Terrapene carolina," *Copeia* (March 1939): 14–20.

A Florida Picnic

1 Rudyard Kipling, *The Jungle Book* (1894; reprint, Harmondsworth, England: Penguin, 1961). The story of the mugger of Mugger Ghat appears in the chapter titled "The Undertakers." The mugger also said, "Respect the aged!"

All the Way Down upon the Suwannee River

1 William Bartram, *Travels Through North and South Carolina, Georgia, East and West Florida* (Philadelphia: James and Johnson, 1791; New York: Penguin, 1988). See also Francis Harper, *The Travels of William Bartram* (New Haven: Yale University Press, 1958).

2 Archie Carr, *Ulendo: Travels of a Naturalist In and Out of Africa* (New York: Knopf, 1964; Gainesville: University Press of Florida, 1993).

3 Thomas B. Thorson, "Movement of bull sharks, *Carcharhinus leucas,* between Caribbean Sea and Lake Nicaragua demonstrated by tagging," *Copeia* (July 1971): 336–38.

4 Apparently the Suwannee chicken, *Pseudemys concinna suwanniensis,* is holding its own. Paul Moler, biological administrator with the Florida Game and Fresh Water Fish Commission, thinks that as our society becomes more urban the habit of eating off the land decreases. That is good news for the Suwannee chicken, and it probably is a help to the gopher tortoise, though loss of habitat is the main difficulty for the gopher.

5 Ichetucknee Springs State Park continues to be one of the loveliest and most popular places to visit—particularly on a hot summer day. The Park Service has sought a balance between people tubing and the fragile aquatic plants in the spring run. They have found that the level of wear and tear on the plants is acceptable if only 750 people per day are permitted to tube the entire length of the run. In addition, tubing is only allowed June 1 to Labor Day. Canoeing, however, is permitted year round. For more information contact Ichetucknee Springs State Park, Route 2, Box 108, Fort White, FL 32038; telephone: (904) 497-2511.

Suwannee River Sturgeon

1. James A. Huff, "Life history of Gulf of Mexico sturgeon, *Acipenser oxyrhychus desotoi*, in Suwannee River, Florida," *Florida Marine Research Publications* (November 1975): 1–32.

2. It has been difficult to determine where the sturgeon spawn. Since 1985, Stephen Carr, with support from the Phipps Florida Foundation, has diligently monitored the sturgeon each year. In 1988 the Fish and Wildlife Service of the United States Department of the Interior initiated a program designed to restore sturgeon to their former waterways. Stephen places beepers on big female sturgeon, then follows them day and night. He thinks they spawn near spring outflows. Sure enough, in June 1993 eggs were found near the Alapaha Rise far up the Suwannee, and in August sturgeon fry were found there.

An Introduction to the Herpetology of Florida

1. Jacques Le Moyne de Morgues, a Huguenot artist, accompanied René Goulaine de Laudonnière to Florida in 1564–65. Laudonnière's narrative of the events in Florida, illustrated with forty-two engraved reproductions of the drawings Le Moyne made while in Florida, were published in Germany after Le Moyne's death. (The translator changed the name Jacques to Jacob.) These are the earliest known pictures of Indians of North America. On how the expedition came about see David I. Bushnell, Jr., "Drawing by Jacques Le Moyne de Morgues of Satirioua, a Timucua chief in Florida, 1564," Smithsonian Miscellaneous Collections, vol. 81, no. 84 (1928). According to Bushnell, the English translation of Le Moyne is as follows:
   > Charles IX, King of France, having been notified by the Admiral de Chatillon that there was too much delay in sending forward the re-enforcements, needed by the small body of French whom Jean Ribaud had left to maintain the French dominion in Florida, gave orders to the admiral to fit out such a fleet as was required for the purpose. The admiral, in the mean while, recommended to the king a nobleman of the name of Renaud de Laudonnière; a person well known at court, and of varied abilities, though experienced not so much in military as in naval affairs. The king accordingly appointed him his own lieutenant, and appropriated for the expedition the sum of a hundred thousand francs. . . . I also received orders to join the expedition, and to report to M. de Laudonnière. . . . I asked for some positive statements of his own views, and of the particular object which the king desired to obtain in commanding my services. Upon this he promised that no services except honorable ones should be required of me; and he informed me that my special duty, when we should reach the Indies, would be to map the seacoast, and lay down the position of towns, the depth and course of rivers, and the harbors; and to represent also the dwellings of the natives, and whatever in the province might seem worthy of observation: all of which I performed to the best of my ability, as I showed his majesty, when, after having escaped from the remarkable perfidies and atrocious cruelties of the Spaniards, I returned to France."

2. William Bartram, *Travels Through North and South Carolina, Georgia, East and West Florida* (Philadelphia: James and Johnson, 1791; New York: Penguin, 1988). See

also Francis Harper, *The Travels of William Bartram* (New Haven: Yale University Press, 1958).

3 Edwin D. Cope, "On the snakes of Florida," *Proceedings of the United States National Museum* (1888): 381–94.

4 Einar Loennberg, "Notes on reptiles and batrachians collected in Florida in 1892 and 1893," *Proceedings of the United States National Museum* (1894): 317–39.

5 Clement S. Brimley, "Records of some reptiles and batrachians from the southeastern United States," *Proceedings of the Biological Society of Washington* (1910): 8–18.

6 The Fall Line is where the coastal plain meets the Piedmont upland of the Appalachian highlands. Where streams that traverse the older and harder rocks of the uplands enter the younger sediments of the coastal plain, falls or rapids occur. The Fall Line was of enormous importance in the development of the eastern seaboard because the falls in the rivers marked the upper end of uninterrupted navigation and also provided water power for mills. Looking south from I-10, east of Tallahassee, one can easily see the Fall Line as a marked dip in the terrain. The white-water rapids of the Suwannee River near Ellaville are another indication of the Fall Line in Florida.

7 Charles Schuchert, *Historical Geology of the Antillean-Caribbean Region* (New York: Riley, 1935).

8 E. P. St. John, "Rare ferns of central Florida," *American Fern Journal*, 1936, no. 26: 41–55.

9 This refers to personal conversations held in the 1940s with Dr. T. H. Hubbell and Sidney Stubbs of the University of Florida biology department.

Florida Vignettes

1 It is thought that the gopher tortoise may no longer exist outside of protected areas by the year 2000. That's a shame and a needless loss of one of the state's most attractive, mild, and amenable animals.

Some steps are being taken. The state has designated the gopher tortoise as a "species of special concern," and methods are being tested and evaluated for relocating gopher colonies caught in the path of construction. Although it takes a little effort, gophers tortoises can live in close proximity to development, according to Elizabeth Knizley, secretary of the Gopher Tortoise Council (formed in 1978) and an effective friend of this gentle animal. She tells me that "innovative land developers are setting aside areas for gophers within communities. When communities cannot set aside adequate areas, potential restocking sites, such as reclaimed mining lands and public water supply wellfields, may possibly serve as home sites for displaced gophers. Beyond relocation, increased protection at the legislative level, protection of gopher habitat in the form of preserves, and public education are required to reverse the gopher's alarming decline."

To help the gopher contact the Gopher Tortoise Council, Florida Museum of Natural History, University of Florida, Gainesville, FL 32611.

2 James E. Lloyd, "Aggressive mimicry in *Photuris:* Firefly femmes fatales," *Science*, 1965, no. 149: 653–54.

3 Archie F. Carr, "A key to the breeding songs of the Florida frogs," *Florida Naturalist*, 1934, no. 7: 19–23.

Tails of Lizards

1 Archie Fairly Carr, Jr., *A Contribution to the Herpetology of Florida,* Biological Science Series, no. 3 (Gainesville: University of Florida, 1940), 1–118.

Alligator Country

1 Since 1969, when Archie wrote about alligators, both the alligator and the Florida Game and Fresh Water Fish Commission have made great strides. The alligator responded quickly to protection and can once again be found in wetlands throughout the state. The commission has, since 1988, been carrying out a careful Alligator Management Program that is based on the premise that the economic value derived from alligator hides, meat, and eggs can provide economic incentives for conserving alligators and preserving their wetland habitat.

A large percentage of Florida's wetlands are privately owned, and the commission has offered these owners an incentive for maintaining these habitats by providing them an opportunity to manage and harvest alligators from their lands. In 1992 there were seventy-three properties comprising more than three hundred thousand acres registered in this program. Each property is reviewed annually by Alligator Management Program biologists and assigned a conservative harvest quota. A total of 875 alligators were harvested on private lands in 1992.

Harvesting alligators on public lands is another component of the Alligator Management Program. The commission believes that the hunt attracts "statewide, national, and international interest and provides an excellent opportunity to inform the public about the value of alligators and wetlands, while allowing participants to benefit from this renewable resource." In 1992, 176 people were chosen by random drawing out of more than twelve thousand permit applications. Of these, 143 completed the training and orientation programs and were issued harvest permits and fifteen alligator harvest tags each. 1,491 alligators—70 percent of the number of tags issued—were harvested.

Since Archie wrote *Alligator Country* nearly twenty-five years ago, many advances have been made in alligator farming. Today you can even buy high-protein alligator pellets from Purina. According to Dr. James Perran Ross, executive officer of the Crocodile Specialist Group of the International Union for the Conservation of Nature and assistant scientist at the Florida Museum of Natural History, Gainesville, "Alligator farming and ranching in Florida reached a peak in mid-1990, and since that time a decline in world prices for skins has caused a contraction in the industry. In 1992 there were fifty-five licensed alligator farms in Florida. These are holding about one hundred thousand captive alligators at any one time and producing twenty-five thousand to thirty-five thousand skins and more than two hundred thousand pounds of meat for sale each year. Total annual value of sales is in the vicinity of $3.5 million a year (down from about $5 million in 1990). About a quarter of the captive alligators are born on farms of captive parents, and the remainder are collected on a controlled, sustainable basis from wild alligators. Taxes and fees from this program support the Alligator Management Program of the Florida Game and Fresh Water Fish Commission, which regulates the collection of alligator eggs and hatchlings for ranches, a controlled hunt of wild alligators on both state-owned and private lands, and a nuisance alligator removal program available from the commission.

"Under this regime of controlled use, alligator populations in the state are stable or growing, though loss of habitat to urban development is a continuing problem.

"The Florida alligator program is dwarfed by a similar program in Louisiana, where more than 130 farms produce sixty thousand to seventy thousand skins annually. Altogether the U.S. alligator industry produces more than one-third of the legal crocodile skins in trade worldwide. The majority of these skins are exported to Italy, France, and Japan. Meat consumption is evenly divided between domestic sales and exports, largely to Taiwan."

The Alligator Management Program also monitors statewide alligator population trends and sponsors programs to educate the public about the important role the alligator plays in Florida's wetland ecosystems. The educational programs are supplemented by the distribution of brochures about the life history and biology of alligators and how Floridians can safely coexist with them.

A flaw in this elaborate and meticulously carried out program is the emphasis on the monetary value of hides, meat, and "sport hunting" permits in defining the value of alligators. In an effort to increase the tangible—monetary—value of alligators, the commission in 1993 encouraged sport hunting of alligators. Hunters armed with .357 magnum spears were chasing around in shallow lakes at night in airboats! I think the whole approach is wrong. The alligator is not here to provide meat, skin, and joy of hunting for man. That arrogant assumption is out of date. Today we recognize the important role the alligator plays in molding and maintaining the wetland landscapes in the South. Living in harmony with this big predator will take some doing. But our brains are bigger than the gators' are, and we should be able to come up with a plan to accommodate both people and alligators in Florida. If it appears that the gator population needs to be reduced in any given lake, the game commission personnel should carefully assess the problem. If necessary, let them kill the extra gators and sell off the hides and meat. If there is no market for these items, let them bury the alligators. There is no exucuse for sport hunting. I also think it would help if the commission dropped the word *resource* and called the animals *alligators* instead. For more information contact Alligator Management Program, Wildlife Research Laboratory, Florida Game and Fresh Water Fish Commission, 4005 South Main Street, Gainesville, FL 32601.

2 Jacob Le Moyne, "Indorum Floridam provinciam inhabitantium eicones," in Theodore Bry, ed., *Voyages en Virginie et en Floride* (Liège, 1591; reprint, Paris: Ducharte et Van Buggenhoudt, 1926). See also "An Introduction to the Herpetology of Florida," n. 1.

3 William Bartram, *Travels Through North and South Carolina, Georgia, East and West Florida* (Philadelphia: James and Johnson, 1791; New York: Penguin, 1988). In the 1950s Francis Harper retraced many of Bartram's travels. In 1958 he published a naturalist's edition of *The Travels of William Bartram* (New Haven: Yale University Press, 1958), edited with commentary and an annotated index. While we luxuriated in referring to our own copy of the second edition of Bartram it was Francis Harper's edition that we depended on. I find our copy filled with little pieces of paper with Archie's notes on them. It's a grand book and a proper tribute to that marvelous naturalist William Bartram. A side note: The zoology department of the University of Florida is housed in two connected buildings; one is William Bartram Hall, and the other is Archie Carr Hall. This gives me great pleasure.

4 William Bartram, *Travels Through North and South Carolina, Georgia, East and West Florida* (Philadelphia: James and Johnson, 1791; New York: Penguin, 1988). See also Francis Harper, *The Travels of William Bartram* (New Haven: Yale University Press, 1958).

5 R. H. Chabrek, "The movements of alligators in Louisiana," *Proceedings of the Southeast Association of Game and Fish Commissions, 1965* (1966): 102–10.

6 Edward Avery McIlhenny (1872–1949) made an enormous contribution to the understanding of the behavior of the alligator. McIlhenny lived on Avery Island, Louisiana, which, according to McIlhenny, "is a series of hills rising about two hundred feet above the coastal plain of south Louisiana and is located about halfway between New Orleans and the Texas line." Avery Island is where that marvelous and unique McIlhenny's Tabasco Sauce is produced. It also happens to be the center of the greatest abundance of Louisiana alligators. McIlhenny, a layman, made meticulous observations on the engaging behavior of alligators and reported his findings in a charming little book, *The Alligator's Life History* (Boston: Christopher Publishing House, 1935). Before his death he was instrumental in establishing more than 175,000 acres of wildlife sanctuary in the marshlands of southern Louisiana. In 1976 McIlhenny's book on the alligator was republished by the Society for the Study of Amphibians and Reptiles, Miscellaneous Publications, Facsimile Reprints in Herpetology, with a foreword by Archie Carr and an index to recent literature by Jeffrey W. Lang. In summing up McIlhenny's contribution to understanding the alligator, Archie states: "This patient and redoubtable man contributed vastly in the only ways we know to save beleaguered wild species: by studying their biology and by setting aside wild landscapes in which they are safe from human depredations."

A Subjective Key to the Fishes of Alachua County, Florida

1 In the fall of 1936, when I first met Archie, I had a job as a wildlife technician with the Resettlement Administration in Welaka, Florida. One of my tasks was to identify the fish of the area. Archie gave me this key and assured me it would work. It did, indeed—but I wept with frustration while using it. In 1941 the American Society of Ichthyologists and Herpetologists (ASIH) held its annual meeting in Gainesville. The key was printed and distributed at that time as an issue of *Dopeia* (the name of the journal of the ASIH is *Copeia),* published by "The American Society of Fish Prevaricators and Reptile Fabricators." I want to thank Dr. Brooks Burr, editor of *Copeia* (1993), for permission to reprint the key.

Carnivorous Plants

1 Durland Fish and Donald Hall, "Succession and stratification of aquatic insects inhabiting the leaves of the insectivorous pitcher plant, *Sarracenia purpurea,"* American Midland Naturalist, 1978, no. 99: 172–83.

2 Durland Fish, "Structure and composition of the aquatic invertebrate community inhabiting epiphytic bromeliads in South Florida and the discovery of an insectivorous bromeliad" (Ph.D. Diss., University of Florida, Gainesville, 1976).

The Moss Forest

1 Tony Jensen was a most observant man who cared a great deal about the forests in Florida. Tony, who died in 1985, was a neighbor of ours in Micanopy. He and Archie had many conversations about trees in general and live oaks in particular, and Archie put great store in what Tony had to say about the growth rate and age of live oaks.

2. San Felasco Hammock State Preserve was one of the first areas purchased by the state of Florida with the Environmentally Endangered Land Fund in the early 1970s. It is made up of sixty-five hundred acres. There are hiking and horseback trails, and the preserve is open to the public year round. For more information contact San Felasco Hammock State Preserve, Devil's Millhopper State Geological Site, 4732 Millhopper Road, Gainesville, FL 32601; telephone, (904) 336-2008.

3. Archie Carr, *Ulendo: Travels of a Naturalist In and Out of Africa* (New York: Knopf, 1964; Gainesville: University Press of Florida, 1993), xix.

Water Hyacinths

1. James O'Hara, "Invertebrates found in water hyacinth mats," *Quarterly Journal of the Florida Academy of Science,* 1967, no. 30: 73–80.

2. Florida has finally come to terms with water hyacinths. No longer does it wait until a lake or river is clogged with plants and then spray with massive doses of herbicides. Today the objective is to maintain hyacinths at the lowest possible level, knocking them back with spray before they get out of hand. This policy has reduced the amount of spraying—and the cost. In addition, three insects have been introduced that like to eat hyacinths. In 1972 two weevils, *Neochetina bruchi* and *Neochetina eichhorniae,* were released; in 1977 a moth, *Sameodes albiguttalis,* was added to the arsenal. After about ten years the three insects are well established, and though their effectivness is hard to assess, they must play an important role in the management of hyacinths.

Triple-Clutchers

1. Archie Carr, *Handbook of Turtles: The Turtles of the United States, Canada, and Baja California* (Ithaca, N.Y.: Cornell University Press, 1952).

2. Peter C. H. Pritchard, *Encyclopedia of Turtles* (Neptune, N.J.: T. F. H. Publishing, 1979).

A Dubious Future

1. After some years of pushing and pulling, the University of Florida realized that Lake Alice was indeed a solid asset, and today it remains a beautiful little wild lake on the campus. For a while, there were fears regarding too-aggressive alligators, but the construction of two small islands for the alligators to bask on and the posting of signs prohibiting their feeding has resulted in a mutually beneficial relationship.

2. For more than three decades the Caribbean Conservation Corporation (CCC) has carried out a green turtle tagging program at Tortuguero. The program is the longest-running study of its kind in the world. Since 1987 Archie's work has been continued by the CCC, whose executive director is David Carr, and by the Archie Carr Center for Sea Turtle Research at the University of Florida, whose director is Dr. Karen Bjorndal, one of Archie's former graduate students. The two groups coordinate the International Cooperative Tagging Project, which assists tagging efforts of sea turtles throughout the world.

In Costa Rica, the CCC has implemented a comprehensive zoning plan for the last remaining stretch of unprotected rain forest in order to protect the sea turtle rookery at Tortuguero. To educate and interest the forty-seven thousand visitors who now come to Tortuguero each year, CCC is in the final stages of building an environmental interpretation and extension center. The center includes educational exhibits on sea turtle biology and the coastal environment, a training program for tour guides to the nesting beach, and a program to involve the local community in ecotourism. Local folks who once hunted turtles now guide a growing number of ecotourists along the nesting beach.

In Nicaragua, the CCC is developing a management plan for the five-thousand-square-mile Miskito Coast Protected Area, one of the largest resident green turtle feeding grounds in the world. This magnificent, essentially untouched area contains some of the richest coastal and marine environments anywhere in the hemisphere. It supports a diverse abundance of aquatic and terrestrial wildlife, including manatees, dolphins, and rare water birds. An essential part of the plan is the integration of environmental and economic interests and the promotion of cooperation between indigenous people and the national government.

In Central America, the CCC is involved with the Wildlife Conservation Society (formerly the New York Zoological Society) in a bold regional conservation initiative called Paseo Pantera—"path of the panther." Archie Carr III (often called Chuck), regional coordinator of Mesoamerican Programs for the society, developed the idea for this elegant project, which strives to preserve biological diversity and enhance wildlands management on the Central American isthmus by linking a chain of parks and protected areas across seven countries from Guatemala to Panama, creating a protected corridor connecting North and South America. The goal of Paseo Pantera is to provide the methods, tools, and knowledge for the nations of Central America to work together toward conservation. The seven presidents of the region already have signed a Central American Biodiversity Treaty calling for the corridor.

In Florida, the CCC has been instrumental in helping to establish the country's first national preserve for sea turtles—the Archie Carr National Wildlife Refuge. This twenty-mile stretch of beach near Melbourne is the nesting site for the largest population of loggerhead turtles in the Western Hemisphere (about sixty thousand) and the largest green turtle rookery in the United States (about one thousand).

Through a new program, the Sea Turtle Survival League, the CCC is working to implement stricter government regulations to protect U.S. sea turtle populations, including a ban on certain commercial fishing activities, limits on coastal development and pollution, and beach lighting ordinances. The group is also working to raise public awareness of sea turtle conservation issues.

For more information about sea turtle conservation or the Paseo Pantera contact the Caribbean Conservation Corporation, P.O. Box 2866, Gainesville, FL 32602-2866; telephone, (800) 678-7853.

Eden Changes

1 John Kunkel Small, *From Eden to Sahara: Florida's Tragedy* (Lancaster, Pa.: Science Press, 1929); and Thomas Barbour, *That Vanishing Eden: A Naturalist's Florida* (Boston: Little, Brown, 1944).

2 Jacob Rhett Motte, *Journey into Wilderness: An Army Surgeon's Account of Life in*

Camp and Field During the Creek and Seminole Wars, 1836–1838, James F. Sunderman, ed. (Gainesville: University of Florida Press, 1953).

3 For an explanation of cut-bait fishing, please see "All the Way Down Upon the Suwannee River," p. 69.

4 Marjorie Kinnan Rawlings, *The Yearling* (1938; reprint, New York: Macmillan, 1986).

5 Florida has continued a vigorous program of setting aside unique landscapes before they are developed or priced out of the reach of public purchase. In 1979 the Conservation and Recreation Lands Act (CARL) was passed with funding from mineral extraction severance taxes amounting to $15 million to $40 million annually. Save Our Coasts, a $275 million bond program, was passed in 1981. In the same year a program called Save Our Rivers, under the aegis of the Water Management Districts and funded by documentary stamps amounting to $30 million to $40 million annually, was approved by the voters in Florida. In 1987 funding for the CARL program was increased by bonding. Realizing that time to purchase prime landscapes was running out, the state in 1990 passed the Preservation 2000 Act, which provides $300 million for land purchase in the next ten years. In response to public demand, Florida has one of the finest public lands programs in the nation. For information concerning Florida's magnificent state parks, contact the Florida Department of Environmental Protection, Division of Recreation and Parks, 3900 Commonwealth Boulevard, Tallahassee, FL 32399; telephone, (904) 488–9872.

Index

Adams, Neal, 145–47
Alachua: named by early Indians, 52
Alachua Lake: wet-dry cycles, 4; steamboats on, 14
Alapaha Rise (spring on Suwannee River), 56
Allen, Ross: on alligators 112, 162; on turtles, 220
Alligator, American *(Alligator mississippiensis)*: behavior in Wewa Pond, 6, 9, 159; Bartram's description of, 32, 33; in Homosassa Springs, 34; at vulture feeder, 48, 49; in Billy's Lake, 56; in Suwannee River, 72; threats to its existence, 104; range of, 105; feeding habits, 105–7; physical description of, 110, 111; predation of young, 112, 113; homing urge, 113, 114; courting, 114, 115; nesting, 115, 116, 117; relation with herons, 121; voice of, 122, 123; need to keep in landscape, 123, 124; attempted capture of, 160–64; population growth, 237
Alligator, Chinese *(Alligator sinensis)*, 105
Alligator farms, 249
Alligator Management Program (Florida), 249, 250
Antillean fauna, 86
Apalachee Indians, 52
Apalachicola ravines, 86
Apalachicola River: bull shark in, 66; commercial sturgeon fishing, 74, 75; influence on flora and fauna, 86
Archie Carr Center for Sea Turtle Research, 252
Archie Carr National Wildlife Refuge, 253
Archosaurs, 104

Armadillo, nine-banded *(Dasypus novemcinctus)*: as eagle fodder, 47–49; reintroduction to Florida, 206, 207; depredations by, 207–9
Arrowhead *(Sagittaria lancifolia)*, 198
Artesian springs, 62
Audubon, John James: account of live oak harvesting, 173
Audubon Society: halting plume hunting, 238

Barbour, Thomas, 232; purchase of Thomas Farm Dig, 240
Bartram, William: influence on Samuel Coleridge, 15; account of jubilee, 32, 33; account of Indian ocean trips, 52, 53; description of Suwannee, 57, 58; description of Manatee Springs, 59, 60; reference to reptiles, 84, 85; reaction to alligators, 109, 110; on springs, 239
Bass, largemouth *(Micropterus salmoides)*: in jubilees, 32, 34, 37; description of, 134
Bass, smallmouth *(Micropterus dolomieu)*, 134
Bass, Suwannee *(Micropterus notius)*, 61, 65
Basswood *(Tilia caroliniana)*, 196
Bay, loblolly *(Gordonia lasianthus)*, 201
Bay, shore *(Tamala littoralis)*, 226
Bay, sweet *(Magnolia virginiana)*, 201
Bayheads, 201
Beaver *(Castor canadensis)*, 237
Beech, blue *(Carpinus caroliniana)*, 86
Begonia *(Begonia cucullata)*, 202
Bellamy, Buck, 226
"Ben" (dog): and armadillos, 47
Berlese funnel, 207–8

255

Big Pine Key: flatwoods in, 194
Billy's Lake, 56
Birch, river *(Betula nigra)*, 64
Bison: on Paynes Prairie in Pleistocene, 19
Bivens Arm, 216, 217
Bjorndal, Karen, 252
Blackfly larvae, 64
Bladderwort *(Utricularia spp.)*, 149, 150, 199
Blountstown, Fla.: and sturgeon, 75
Bobcat *(Felis rufus)*, 12, 195, 206
Boleck (Billy Bowlegs): attacked by Andrew Jackson *(1818)*, 53
Bonnet *(Nuphar luteum)*, 200, 214
Bougard, Russell: bull shark encounter, 66–67
Bowfin *(Amia calva)*, 15
Brachiopod *(Glottidia pyramidata)*: eaten by sturgeon, 78
Braren, Herbert, 226
Bream, bluegill *(Lepomis macrochirus purpurescens)*, 34, 134
Bream, redbreast *(Lepomis auritus)*, 56
Bream, speckled *(Sclerotis p. punctatus)*, 135
Breckenridge, Henry Marie (judge), 174
Bromeliad *(Catopis berteroniana)*, 151
Brothers River: bull shark caught in, 66
Bufo marinus, 98
Bull shark *(Carcharinus leucas)*, 65, 66–67
Bullhead, blue *(Ictalurus catus)*, 127
Bullhead, flatheaded *(Ameiurus platycephalus)*, 128
Bullhead, marbled *(Ictaluras nebulosus maramoratus)*, 8, 9, 128
Bullhead, yellow *(Ictalurus natalis)*: in underground waters, 5
Butterwort *(Pinguicula spp.)*, 149, 150
Buttonwood *(Conocarpus erectus)*, 201

Cabomba, 203
Caddis fly larvae, 64
Caldwell, John M., 54
Caloosahatchee epoch (of Pliocene), 88
Calos, Bay of, 53
Camels (Pleistocene), 19, 62
Caribbean Conservation Corporation: work with sturgeon, 74; founding of, 234; program today, 252–53
Carpinus caroliniana, 196
Carr, Archie: biographical sketch, xiii–xv
Carr, Archie, III (Chuck), 69, 253
Carr, David, 67, 252
Carr, Marjorie Harris, 70, 71
Carr, Stephen, 69, 80–87
Carr, Tom (brother to Archie), 226
Carr, Tom (son of Archie): and bull shark, 66
Carson, Rachel, 187
Catfish, gaff-topsail *(Bagre marinus)*, 34, 36
Catfish, sea *(Galeichthuys felis)*, 34
Catfish, southern channel *(Ictalurus punctatus)*, 126
Cattail *(Typha spp.)*, 199
Cattle industry (Florida), 17
Cavally-jack *(Caranx hippos)*, 36
Caves: crustacea inhabiting, 4
Cedar Key, Fla., 54, 55
Cedar, red *(Juniperus silicicolas)*, 59
Cedar, stinking *(Tumion taxifolium)*, 86
Chabreck, Robert, 113
Chara ponds, 198
Chassahowitzka Spring, 63
Chattahoochee River: species migration, 85
Cocoa-plum *(Chrysobalanus icaco)*, 201
Colubridae (snake family): distribution of, 157
Conservation (wilderness), 233; value to human spirit, 230–31
Conservation and Recreation Lands Act (CARL), 254
Coontail *(Ceratophyllum demersum)*, 61, 199
Coontie *(Zamia floridana)*, 194
Cooter, Florida *(Pseudemys f. floridana)*, 222–23
Cooter, peninsula *(Pseudemys floridana peninsularis)*, 220–21
Cooter, red-bellied *(Pseudemys nelsoni)*, 37
Cooter, Suwannee *(Pseudemys concinna suwanniensis)*, 223
Cordgrass *(Spartina spp.)*, 200
Corkwood *(Leitneria floridiana)*, 59
Cornell University: research on Okefenokee, 56
Costa Rica, 234
Crab, blue *(Calinectes sapidus)*: in Daphne jubilee, 24; in St. Johns River, 37–38
Crane, sandhill *(Grus canadensis)*, 16
Crawl-a-bottom *(Hadropterus nigrofasciatus)*, 135, 136

Crayfish *(Procambarus fallax)*: and Wacahoota jubilee, 26, 27
Crayfish, cave *(Troglocambarus maclini)*, 4
Creeks (Indians): settled on Suwannee River, 52
Croaker *(Micropogonias undulatus)*, 37–38
Crocodile, American *(Crocodylus acutus)*: occurrence in Florida, 86
Crow, fish *(Corvus ossifragus)*, 48, 221–22
Crystal Beach: submarine springs in Gulf of Mexico, 36
Crystal Spring, Fla., 63
Cuba: early Indian trade to, 52–53
Cuban tree frog *(Hyla septentrionalis)*: introduction to Florida, 86
Cut-bait fishing, 240, 241
Cypress, bald *(Taxodium distichum)*: used for dugout canoes, 52; harvesting of, 58–59; host to Spanish moss, 167, 200
Cypress, pond *(Taxodium ascendens)*, 198

Damselflies, 64
Daphne, Ala.: jubilees, 22–25
Darter *(Etheostoma spp.)*, 64
Darter, brown *(Villora edwini)*, 135–38
Darter, swamp *(Hololepis barratti)*, 135–38
Darwin, Charles: interest in insectivorous plants, 148, 149
De Soto, Hernando: and Battle of Mapitaca, 52
Deer, white-tailed *(Odocileus virginianus)*, 195
Demory Hill, 59, 73
Devonian period: brachiopods and sturgeon, 83
Dickinson, Jonathan: on use of Spanish moss, 168
Dog hobble *(Leucothoe axillaris)*, 201
Dogwood *(Cornus florida)*, 196
Dogwood, Jamaica *(Piscidia piscipula)*, 197
Dowling Park (on Suwannee River), 80
Downs, Shelly: and attempted alligator capture, 161–63
Dragonflies, 64
Duck, Florida *(Anas fulvigula)*, 31
Duck, ring-neck *(Aythya collaris)*, 10
Duck, wood *(Aix sponsa)*, 178–79
Ducks, migratory: loss of, 8–11
Duckweeds *(Lemna, Spirodela, Wolffia,* and *Wolffiella)*, 11, 198

Eagle, bald *(Haliaeetus leucocephalus)*, 48
East Pass (on Suwannee River), 59
Ecological succession: in northern Florida, 171
Eel *(Anguilla bostoniensis)*, 126
Eel grass *(Sagittaria sp.)*, 61
Egret, cattle *(Bubulcus ibis)*: immigration to Florida, 237
Ellaville, Fla., 51, 54
Emergency scent control, 143–47
Endemism: factors contributing to, 87
Everglades: gator holes, 120; limestone ridges, 194

Faber (pencil factory), 59
Fall line, 248
Fannin Springs, Fla., 54, 58
Fern *(Dryopteris floridanum)*, 202
Fern *(Lorinseria areolata)*, 202
Fernandina Beach, Fla., 55
Ferns (relic): study by E. P. St. John, 88
Fichero, Angelo, 79
Fireflies: reasons for flashing, 96; *Photinus* and *Photuris*, 96
Fish, Durland, 151
Flagfish *(Jordanella floridae)*: in Wacahoota jubilee, 27
Flatwoods ponds, 198
Flatworms, aquatic, 64
Flier *(Centrarchus macropterus)*, 132
Flint River: species migration on, 85
Floating islands: on Orange Lake, Fla., 214
Florida (state of): conflict between humans and nature, 237; hope for, 244
Florida fauna *(1940)*: reptiles and amphibians, 85–90
Florida Game and Fresh Water Fish Commission: management of alligators, 249–50
Florida, University of: student specimin collectors, 210–11, 212
Floridan Aquifer: formation of, 62
Flounder: in Daphne jubilee, 24
Fluctuating ponds, 198
Fluvial swamps, 200–201
Fort Drane, Fla., 25
Fort Fannin, Fla., 54
Fort Micanopy, Fla., 25
Fowler Bluff (on Suwannee River), 67
Fox, gray *(Urocyon cineroargenteus)*, 195

Frog, green tree *(Hyla cinerea)*, 98
Frog, greenhouse *(Eleutherodactylus ricordii)*: introduction to Florida, 86
Frog songs: subjective key, 97; spring songs at Wewa, 98–99

Gall berry *(Ilex glabra)*, 194
Gallinule, Florida *(Gallinula chloropus cachinnans)*, 163–64
Gar, longnose *(Lepisosteus osseus)*, 34, 57, 126
Gar, short-nose *(Lepisosteus platyrhincus)*, 126
Gator fleas *(Asellus hobbsi* and *Crangonyx hobbsi)*, 4
Gator holes: importance of, 117–18
Gecko reef *(Spherodactylus notatus)*, 86
Ghiselin, Margaret Odlund, 74
Gilchrist County, Fla. (Thomas Farm Dig), 240
Giovannoli, Leonard, 63
Glowworms: at Wewa Pond, 95–96
Glyptodont (Pleistocene): on Paynes Prairie, 19, 205
Goniabasis (snail), 64
Gopher, pocket *(Geomys pinetis)*, 195
Gopher Tortoise Council, 248
Grasshoppers *(Melanosphus spp.)*, 195
Grasshoppers: and planted pastures, 18, 19
Grassland fauna: persistence of, 20
Grinnels, 32
Gulf Coastal Lowlands, 57
Gulf of Mexico: and Suwannee River, 55; offshore springs, 63; bull sharks' breeding place, 66
Gum, black *(Nyssa biflora)*, 194
Gum, sweet *(Liquidamber styraciflua)*, 196

Hacienda La Chua, 52
Hackberry *(Celtis laevigata)*, 167, 196
Halifax River, 38
Hall, David, 151
Hammock: live oak, 169; defined and described, 171; hydric, 171; xeric, 171; low, 196; mesophytic, 196; shallow, 196; tropical, 196; upland, 196; ponds, 198–99
Harper, Francis: retracing William Bartram's route, 33
Hart Springs (on Suwannee River), 81
He-huckleberry *(Cyrilla racemiflora)*, 201

Hellgrammites, 64
Heron, black-crowned night *(Nycticorax n. hoactlii)*, 227
Heron, great blue *(Ardea herodius)*, 8–9, 227
Heron, little blue *(Florida caerulea)*, 6–8, 227
Heron, little green *(Butorides v. virescens)*, 227
Heron, Louisiana *(Hydranassa tricolor ruficollis)*, 227
Heron, snowy *(Egretta thula)*: feeding, 14, 18, 19; on Snake Key, 227
Heron, Ward's. *See* Heron, great blue
Herring *(Alosa spp.)*, 37
High pine: as habitat, 89; description of, 195
Hobbs, Horton, 4
Hogchoker *(Trinectes maculatus)*, 65
Holly: American *(Ilex opaca)*, 196; dahoon *(Ilex cassine)*, 201
Holmes Creek (on Suwannee River), 66
Holton Springs (on Suwannee River), 56
Homosassa jubilee, 35–37
Homosassa River, 33
Homosassa Spring, 33, 63, 245
Hornbeam, hop *(Ostrya virginiana)*, 196
Horses (Pleistocene), 19, 62
Hot-bug *(Pelocorus sp.)*, 211
Hubbell, Theodore H., xiii; 88
Hurricane Audrey, 170
Hyacinth, water *(Eichhornia crassipes)*, 200; rolling, 210; ecology of, 212; attempts to control, 219, 252
Hydric hammock. *See* Hammock
Hyla septentrionalis. See Cuban tree frog

Ichetucknee Springs Run, 57, 60–61
Indian Claims Commission, 175. *See also* Seminole Indians
Indian River, 38

Jack-in-the-pulpit *(Arisaema acuminatum)*, 202
Jackson, Andrew, 53, 174
Jackson, Crawford, 223
Jacksonville, Fla., 38
"Jasper" (alligator snapper), 68, 69
Jay, Florida *(Aphelocoma caerulescens)*, 195
Jennings, William, 184

Jensen, Tony, 169, 185
Jim Woodruff Dam: and sturgeon, 75, 82
Jody's Spring: described, 242–43
Jubilee: derivation of word, 22. *See also* Daphne, Ala.; Homosassa jubilee; Wacahoota jubilee

Karst topography: effect of, 3–4; and springs, 239
Kephart, George S., 175
Key West, Fla.: derivation of herpetological fauna, 90
Kilby, John D., 226
Killifish *(Fundulus spp.)*, 26
Killifish, least *(Heterandria formosa)*, 131
Killifish, ocellated *(Leptolucania ommata)*, 132
Killifish, red-finned *(Chriopeops goodei)*, 132
Killifish, Seminole *(Fundulus seminolis)*, 132
Kipling, Rudyard, 49–50
Kissimmee Prairie, 38
Knizley, Elizabeth, 248

Lake Alice: in May, 217; damage to, 218; alligator nest, 230; management today, 252
Lake Dexter (on St. Johns River): Bartram's reference to, 32
Lake George (on St. Johns River), 38
Lake Jackson: draining of, 4
Lake Nicaragua: and bull sharks, 66
Lakes: description of, 199
Lanier, Sidney, 170
Laudonnière, René Goulaine de, 164–65, 247
Le Moyne, Jacques: description of alligator (1591), 84; drawing of alligator, 107–8; observed use of Spanish moss, 167–68; on Laudonnière, 247
Ledwith Lake, Fla., 26–29 passim
Leeches, red-bellied, 27
Lettuce, water *(Pistia striates)*: compared with hyacinth, 213–14
Liche, Carle: on man-eating tree, 151–52
Lichens ("reindeer moss"), 195
Lignum vitae *(Guaiacum sanctum)*, 197
Lignum Vitae Key: tropical hammock on, 197
Lily, water *(Castilia odorata)*, 200

Lime, ogeechee *(Nyssa ogeche)*, 200
Limestone flatwoods: description of, 194
Liverwort *(Riccia spp.)*, 199
Lizard poisoning, 102, 103
Lizard, Florida scrub *(Sceloperus woodi)*, 89, 90
Lizard, Florida worm *(Rhineura floridana)*, 156
Lizard's tail *(Saururus cernuus)*, 202
Lizards: characteristics of, 154–55
Llamas (Pleistocene): on Paynes Prairie, 19
Lloyd, James, 96
Low hammock. *See* Hammock
Ludwigia, 202
Lugwigiantha, 202
Lycosa ceratiola (type of spider), 195

McIlhenny, Edward Avery: on alligators, 116–17; Avery Island and alligators, 251
Mack, Tut, 31
Madison (steamboat), 54–55
Madison, Fla., 53
Madtom *(Schilbeodes leptacanthus* and *s. grinus)*, 64, 128
Magnolia grandiflora (tree), 196
Mahogany *(Swietenia mahogani)*, 197
Maidencane *(Panicum hemitomon)*, 198
Mallard, green-head *(Anas platyrhynchos)*, 31, 164
Mammoth, Columbian (in Suwannee River), 62
Manatee *(Trichechus manatus)*, 236
Manatee Springs, 240; William Bartram's description, 59–60
Mangrove: black *(Avicennia nitida)*, 201; red *(Rhizophora mangle)*, 201, 226
Mangrove swamp, 201
Mapitaca, Battle of, 52
Maple: red *(Acer rubrum)* and sugar *(Acer barbatum)*, 196
Marquez, Tomas Menendez, 52
Marshes: freshwater, 199–200; salt, 200
Mastic *(Mastichodendron foelidissimum)*, 197
Mayflies, 64
Merrimac (ship), 174
Mesophytic hammock. *See* Hammock
Meylan, Anne, and Peter, 222
Minnow, chub *(Notoropis harperi)*: in underground waters, 5

Minnow, eastern star-headed *(Fundulus notti lineolatus)*, 132
Minnow, pug-nosed *(Opsopoedas emiliae)*, 130
Mobile Bay, Ala.: Daphne jubilee, 22–25
Moccasin, cottonmouth *(Ancistrodon piscivoris)*, 158, 225
Monitor (ship), 174
Moore's Prairie, Fla.: Wacahoota jubilee, 26
Mortimer, Jeanne, 71
Mosquito Lagoon, Fla., 38
Mosquito-fish *(Gambusia affinis holbrooki)*, 27, 131
Moss, Spanish *(Tillandsia usneoides)*, 165–68; as a habitat, 184
Motte, Jacob Rhett, 238
Mount, Robert: on skinks, 111
Muck itch, 211
Mud eel, striped *(Pseudobranchus sp.)*, 27
Mud fish *(Amia calva)*, 31, 126
Mud-mary *(Bruneria)*, 198
Mud-midget *(Wolfiella)*, 198
Mulberry *(Morus rubra)*, 196
Mullet *(Mugil spp.)*, 34, 65, 135, 136
Myriophyllum, 203
Myrtle, wax *(Myrica cerifera)*, 91, 201

Narrows of the River (Suwannee River): and canoeing, 55, 56
Narvaez, Panfilo de: crossed Suwannee River (1528), 52
National Science Foundation, 233
Needlefish *(Strongylura marina)*, 65
New River: tributary of Santa Fe, 57
Newnans Lake: moss picking, 168
Newt, Louisiana *(Diemictylus viridescens louisianensis)*, 27
Northern highlands, 57

Oak, blue jack *(Quercus incana)*, 195
Oak, live *(Quercus virginiana)*, 59, 167–71, 178–79; harvesting, described by John James Audubon, 173; as ship timber, 173–75; Santa Rosa Plantation, 174; Indian lawsuit over, 175
Oak, myrtle *(Quercus myrtifolia)*, 195
Oak, sand live *(Quercus geminata)*, 195
Oak, turkey *(Quercus laevis)*, 195
Ocala Island (in Oligocene), 88

Ocala Scrub, 139–47
Ocala uplift, 62
Odlund, Oscar: drift netting for sturgeon, 74
Odlund Island, 73
Odum, Howard, 36
Office of Naval Research, 233
Ogren, Larry, 49
Okefenokee Basin, 55
Okefenokee Swamp, 53, 55
Oklawaha River, 32; tributary of St. Johns River, 63
Old Town, Fla., 240
Oleander, water *(Decadon verticillatus)*, 199
O'Leno State Park, 57
Oligocene: Ocala Island, 88
Olin: tree farming advertisement, 177
Olustee, Battle of, 53
Olustee Creek, 57
Oolite, 194
Opheodrys aestivus (rough green snake), 157
Opuntia cactus: as gopher tortoise food, 94
Orange Creek, Fla., 32
Orange Lake, Fla., 14, 32; floating islands, 214–15
Otters *(Lutra canadensis)*, 27, 72; increase of, 237
Owl, barred *(Strix varia)*, 179

Pachyderms (Pleistocene): on Paynes Prairie, 19
Palatka, Fla., 38
Palm, cabbage *(Sabal palmetto)*, 59, 226
Palm, silver *(Cocothrinax argentea)*, 194
Palmetto, saw *(Serenoa repens)*, 194, 195
Palmetto flatwoods, 194
Panhandle of Florida: herpetological fauna occurrence, 85–86; beavers in, 237
Partington, Bill, and Joan: relationship with indigo snake, 191
Paseo Pantera: description of, 253
Paynes Landing: treaty of, 175
Paynes Prairie, Fla.: wet-dry cycles, 4; description of, 14; snakes, 15, 232; spiders ballooning, 15–16; Pleistocene fauna of, 19, 20; seining experience, 118–20; water hyacinths in, 210

Pearson, Paul, 181–82
Pennywort *(Hydrocotyle ranunculoides)*, 200
Pensacola Navy Yard: and live oaks, 174
Perch, pirate *(Aphredoderus sayanus)*, 132
Perch, red-breasted *(Kenotis megalotis marginatus)*, 135
Perch, speckled *(Pomoxis sparoides)*, 132
Perch, warmouth *(Chaenobryttus gulosus)*, 134
Philotria sp.: and pipefish, 65; habitat of, 202
Phipps, John H. (Ben): and sturgeon in the Apalachicola, 74
Pickerel, bulldog or red-fin *(Esox americanus)*, 27, 57, 130
Pickerel, chain *(Esox niger)*, 130
Pickerelweed *(Pontederia cordata)*, 198
Pine flatwoods, 193
Pine, loblolly *(Pinus taeda)*, 196
Pine, long leaf *(Pinus palustris)*, 193
Pine, pond *(Pinus serotina)*, 194
Pine, sand *(Pinus clausa)*, 195
Pine, slash, 175–77, 193
Pine, South Florida slash *(P. elliotti densa)*, 193
Pine, spruce *(Pinus glabra)*, 196
Pinelands: monitored burning of, 238
Pinfish *(Lagodon rhombides)*, 34
Pinhook Swamp, Fla., 55
Pintail *(Anas acuta)*, 31
Pipefish *(Syngnathus scovelli)*, 37, 65
Pitcher plant *(Sarracenia spp.)*, 149–51
Planarian rapids, 64
Pleistocene: fauna in Florida, 19; interglacial stages, 62; and armadillos, 206
Pliocene islands: habitat for Florida endemics, 88, 89, 90
Plum, pigeon *(Coccoloba diversifolia)*, 197
Poe Springs, Fla., 64
Poisonwood *(Metopium toxiferum)*, 197
Pope, Clifford, 105
Potomogeton, 203
Preservation 2000 Act, 254
Pritchard, Peter, 221
Proserpinaca, 203
Pseudobranchus: in hyacinths, 216
Purslane, water *(Isnardia sp.)*, 61

Racer *(Coluber constrictor)*, 188, 189
Racerunner, six-lined *(Cnemidophorus)*, 155–56
Rails *(Armus sp.)*, 16
Rattler, pygmy *(Sistrurus miliarius barbouri)*: on Wewa farm, 158
Rattlesnake, diamondback *(Crotalus adamanteus)*: on Wewa Farm, 158; warning displays, 189; as a shibboleth for saving wilderness, 234–35
Red tide plague: at Fort Myers, 25
Redbreast (or stumpknocker) *(Lepomis auritus)*, 57, 135
Redfish *(Sciaenops ocellatus)*, 34, 37
Reptiles: defined, 153
"Reynard the Fox" (poem by Masefield), 142
Rheocrene springs, 33, 59; damage to, 243
Rio San Juan, Nicaragua: and bull sharks, 66
Rogers, J. Speed, xiii, 64
Rosemary scrub, 89, 195
Ross, James Perran, 249
Rushes *(Juncus spp.)*, 198
Rushes, black *(Juncus rhoemerianus)*, 59; 200

Sagittaria, 202
Sailfin *(Mollienisia latipinna)*, 131
St. Augustine, Fla.: submarine springs in Atlantic, 36
St. John, E. P., 88
St. Johns River: Bartram's reference to, 32; marine life in, 37, 38; live oak traffic, 173
St. Johns River Water Management District, 245
St. Marks, Fla., 52
St. Marks River, 63
St. Marys River, 55
Salamander, Gulf Coast red *(Pseudotriton montanus flavissimus)*, 202
Salamander, two-lined Southern *(Eurycea bislineata)*, 64
Saltbush *(Baccharis spp.)*, 200, 201
Saltwort *(Batis maritima)*, 200
Samphire: *Salicornia bigelovii* and *Salicornia virginica*, 200
San Felasco Hammock: depredation of, 172
San Felasco State Preserve, 252

Santa Fe Lake, 57
Santa Fe River, 57, 63
Santa Rosa Plantation, 174–75
Saturday Review: author's complaint about advertisement, 177
Save Our Coasts, 254
Save Our Rivers, 254
Sawgrass *(Cladium jamaicensis)*, 59, 200
Sceloporus woodi (scrub pine lizard), 195
Sea cow *(Trichechus manatus)*, 72
Sea Turtle Survival League, 253
Sea-grape *(Coccoloba uvifera)*, 201
Seahorse Island, Fla., 226
Secoffee (leader of dissident Creeks), 52
Seminole Indians: active in Wacahoota Hammock, 25; origin of, 52; reimbursement, 175. *See also* Florida Seminoles
Seminole War (First): action along the Suwannee, 53; treaty of, 175
Shad *(Alosa sp.)*, 37
Shad, gizzard *(Dorosoma cepedianum)*, 128
Shad, Van Hyning's banner *(Signalosa petenis vanhyning's)*, 128
Sheepshead *(Archosargus probatocephalus)*, 34
Shell Landing (on Suwannee River), 66
Shellcracker *(Lepomis microlophus)*: Homosassa jubilee, 37; 135
Sherman, Harley B., xiii
Shiner, big-finned *(Notropis hypselopterus)*, 130
Shiner, coastal *(Notropis roseus)*, 130
Shiner, Florida golden *(Notemigonus chrysoleucus bosci)*, 130
Shiner, iron-colored *(Notropis chalybaeus)*, 130
Shoveler *(Spatula clypeata)*, 9, 10
Shrimp: in Daphne jubilee, 24
Sill (on upper Suwannee River), 55, 56, 57
Silver Springs, Fla.: tributary of Oklawaha River, 63
Silverides, brook *(Labidesthes sicculus vanhyningi)*, 135, 136
Sinkhole ponds, 198
Sinkholes, 3–4
Siren sp.: in Wacahoota jubilee, 26; in hyacinths, 216
Skink, bluetailed *(Eumeces laticeps)*: challenging wren for caterpillar, 100, 101;
reported eating its own tail, 103; care of young, 111
Skink, brown ground *(Scincella lateralis)*, 103
Skink, Florida five-lined *(Eumeces inexpectatus)*, 227
Skink, Florida sand *(Neoseps reynoldsi)*, 89
Skink, red-tailed *(Eumeces onocrepis)*, 89
Skinks: tails, 101, 102, 103; of Wewa Farm *(Scincida)*, 156
Sloths (Pleistocene): on Paynes Prairie, 19
Smith, Homer, 65
Smith, R. C., 223
Snake catching: on Paynes Prairie, 15
Snake, Clark's water *(Natrix clarkii)*, 225
Snake, coral *(Micrurus fulvius)*: on Wewa Farm, 158; eating rat snake, 180
Snake, glass *(Ophisaurus sp.)*, 156
Snake, gopher *(Pituophis)*, 189
Snake, hog-nosed, 190
Snake, indigo *(Drymarchon corais)*, 191
Snake Key, Fla., 226
Snake, pine *(Pituophis sp.)*, 189
Snake, rat *(Elaphe sp.)*, 179, 184
Snake, red-bellied mud *(Farancia abacura)*, 216
Snake, red-bellied water *(Natrix erythrogaster)*, 64
Snake, short-tailed *(Stilosoma extenuatum)*, 89
Snakebird *(Anhinga anhinga)*, 1–2
Snakes: four families of, 156–57; reason for fear of, 187; diet of, 189; mobbed by birds, 189; combat dance of males, 190; as house pets, 191
Snapper, gray or mangrove *(Lutjanus griseus)*, 34, 65
Snook *(Centropomus sp.)*, 34
Sole (hogchoker) *(Trinectes maculatus)*, 37
Soluble limestone, 4
Spartina alternifolia (smooth cordgrass), 200
Spartina spp. (cordgrass), 59
Spatterdock *(Nupha lutea)*: in floating islands, 214
Sphagnum, 202
Spoonbill, roseate *(Ajaia ajaia)*, 72
Spring runs, 202
Springs (Florida): decline of, 239

Squamata: body form of, 155
Stephen Foster State Park, 56
Stingarees *(Dasyatis sabina)*, 37
Stinkjims: mud turtle *(Kinosternon sp.)*, 61; musk turtle *(Sternotherous sp.)*, 216
Stoneflies, 64
Stonewort, 61
Streams: large, 201–2; small, 202
Stubbs, Sidney, 88
Stumpknocker *(Lepomis auritus solis)*, 135
Stumpknocker (speckled bream) *(Lepomis p. punctatus)*, 34, 91, 94
Sturgeon, gulf *(Acipenser oxyrinchus desoti)*: jumping, 66, 67; in Suwannee river, 73–81; loss of fishery, 76; home stream tenacity, 77; tagging, 77–81; stomach contents, 78; growth rate, 78; movements in Suwannee, 78–79; spawning ground search, 80–81; threats to, 82–83; temperament of, 83; advances in research, 247
Sturgeon, short-nosed *(Acipenser brevirostris)*, 125
Sucker *(Erimyzon sucetta)*, 32, 129
Sun Springs, Fla., 81
Sunfish, blue-spotted *(Enneacanthus gloriosus)*, 27, 133
Sunfish, pygmy *(Elassoma evergladei)*: in Wacahoota jubilee, 26, 133
Suwannee (town of), 51, 73; Suwannee Old Town, 53
Suwannee chicken (turtle) *(Pseudemys floridana suwanniensis)*: as food, 67; status today, 246
Suwannee River: route of, 55
Suwannee River Indians: ocean voyages, 52
Suwannee Sound, 55, 59
Suwannee Valley: description of, 51; rheocrene springs in, 59; as trasition zone, 63

Talahasochte, 52, 53, 57
Tampa Bay, Fla.: decline of sturgeon fishery, 76
Tarantina, Sylvester, 78
Taranto, Anthony, 76
Taranto, Joe, 75–76
Tarpon *(Tarpon atlanticus)*, 34, 65, 72
Teal, blue-winged *(Anas discors)*, 9, 10, 12
Tertiary land, 87–88

Thalassia sp. (turtle grass), 68
Thomas, Joseph, 142
Thorson, Thomas: study of bull sharks, 66
Timucua Indians, 52
Titi *(Cyrilla parvifolia* and *Cliftonia monophyla)*, 201
Toad, common *(Bufo terrestris)*, 98
Toad, spadefoot *(Scaphiopus holbrooki)*: habits, 180–81; breeding, 182–83
Topminnow, golden *(Fundulus chrysotus)*, 132; in Wacahoota jubilee, 27
Topminnow, lined *(Fundulus sp.)*, 27
Tortoise, giant (Pleistocene fauna of Paynes Prairie), 19, 62
Tortoise, gopher *(Gopherus polyphemus)*: successful race with author, 95; decline of population, 95; habitat, 195; efforts to protect, 248
Tortuguero, Costa Rica: hyacinth invasion, 216; and green turtle, 233–34
Tree farming, 176
Trillium sp.: in Apalachicola ravines, 86
Tropical hammock. *See* Hammock
Trout, sea *(Cynoscion nebulosus)*, 34
Troy Springs (on Suwannee River), 54, 55
Tucker, James M. (captain of the steamboat *Madison*), 54, 55
Tupelo *(Nyssa aquatica)*, 58, 59
Turtle, alligator snapper *(Macrochelys temmincki)*, 68–72 passim
Turtle, box *(Terrapene carolina)*, 40–46
Turtle, green sea *(Chelonia mydas)*, 233, 252
Turtle, snapping *(Chelydra sp.)*, 27
Turtle, soft-shelled *(Trionyx ferox)*, 216
Turtle, Suwannee chicken *(Deirochelys reticularia)*, 73
Turtles: characteristics of, 153, 154. *See also individual types of turtles*
Tuscawilla, Fla., 2

U.S. Fish and Wildlife Service: fish surveys, 75
Umatilla, Fla., 91
Upland hammock. *See* Hammock

Vallisneria, 202
Van Hyning, George: skink observations, 103
Van Hyning, Oather, 63

Venus's-flytrap *(Dionaea sp.)*, 149
Viperidae: on Wewa farm, 158
Vulture, turkey *(Cathartes aura)*, 48

Wacahoota Hammock, 2, 25
Wacahoota jubilee, 25–29
Wakulla Springs, 63
Wallace, H. K., 91, 226
Warbler, Bachman's *(Vermivora bachmannii)*, 72
Warbler, Parula *(Parula americana)*, 184
Warmouth *(Chaenobryttus gulosus)*, 31, 32, 134
Weeki Wachee Springs, 63
Welika Pond, Battle of, 2
Wewa Pond: naming of, 2–3; description of, 5; natural drainage of, 6; inhabitants, 6–8
White Springs, 54, 55, 57
White, Joseph (Colonel), 174
White, Ted: at Thomas Farm Dig, 240–41

Widgeon *(Mareca americana)*, 12, 31
Wildcat. *See* Bobcat
Wildlife Conservation Society, 253
Willow, black *(Salix nigra)*, 199
Willow, Virginia *(Itea virginica)*, 201
Wire grass. See *Aristida stricta*
Wire grass flatwoods, 194
Withlacoochee (tributary of Suwannee), 57
Woodpecker, ivory-billed *(Campephilus principalis)*, 72
Wortman, Frederic P.: and fish dynamiting (1912), 69–70
Wren, Carolina *(Thryothorus ludovicianus)*: threatened by skink, 100–101
Wright, Albert, and Anna A., xvi

Xeric hammock. *See* Hammock

Yew, Florida *(Taxus floridana)*, 86
Yulee Plantation, 53